BIOLOGY
FUNDAMENTAL PRINCIPLES

THE AUTHORS

Mr. Balaji S. Thorat (Birth 1992) completed B.Sc. (Agri.) from V.N.M.K.V., Parbhani & obtained his Master Degree in Agril. Botany from Dr. B.S.K.K.V., Dapoli. Also he did one year diploma in Agriculture Journalism from Y.C.M.O.U., Nashik. He has throughout a brilliant academic record. His area of specialization is in Genetics & Plant Breeding. B. S. Thorat has six months teaching experience in Genetics & Plant Breeding. He wrote one book on Plant Breeding. He published 14 research papers in various National & International reputed journals and six popular articles to his credits. He contributed 3 book chapters in various books. He has participated, presented his work & 7 abstracts in various National & International conferences, seminars & symposia. He is qualified J.R.F & S.R.F. exams. He also attended Agriculture related training programme which was held by MANAGE at Ratnagiri. B. S. Thorat is a member of several scientific societies and reputed journals. Also he has served as Assistant Professor at GNCA, Mandki-Palvan which is Affiliated to Dr. B.S.K.K.V., Dapoli. Presently, he is perusing Ph.D. at Dr. B.S.K.K.V., Dapoli with specialization Genetics and Plant Breeding.

Dr. Sumit M. Raut (Birth 1987) graduated in Agriculture from M.P.K.V., Rahuri & obtained his M. Sc. & Ph.D. Degree in Agril. Botany from Dr. P.D.K.V., Akola. He has throughout a brilliant academic record. His area of specialization is in Plant Physiology. S. M. Raut is gold medalist in M.Sc. Degree and during Ph.D. programme he got INSPIRE Fellowship. He published 14 research papers in various National & International reputed journals besides, eight popular articles in various magazines. He has published one book chapter. He has participated & published 2 abstracts in various National & International conferences, seminars & symposia. S. M. Raut is a member of several scientific societies. Presently, he is working as a Senior Research Assistant at Dr. B.S.K.K.V., Dapoli on JSW-TPP Mango Project.

BIOLOGY

FUNDAMENTAL PRINCIPLES

– Authors –

Mr. Balaji S. Thorat

Dr. Sumit M. Raut

2023

Associated Publishing Company®

A Division of

Astral International Pvt. Ltd.

New Delhi – 110 002

ISBN: 9789359190990

Publisher's Note:

Every possible effort has been made to ensure that the information contained in this book is accurate at the time of going to press, and the publisher and authors cannot accept responsibility for any errors or omissions, however caused. No responsibility for loss or damage occasioned to any person acting, or refraining from action, as a result of the material in this publication can be accepted by the publisher or the authors. The publisher is not associated with any product or vendor mentioned in the book. The contents of this work are intended to further general scientific research, understanding and discussion only. Readers should consult with a specialist where appropriate.

Every effort has been made to trace the owners of copyright material used in this book, if any. The authors and the publisher will be grateful for any omission brought to their notice for acknowledgement in the future editions of the book.

Published by : **Associated Publishing Company**
A Division of
Astral International Pvt. Ltd.
– ISO 9001:2015 Certified Company –
4736/23, Ansari Road, Darya Ganj
New Delhi-110 002
Ph. 011-43549197, 8130496929
E-mail: info@astralint.com
Website: www.astralint.com

Printed at : **Replika Press Pvt Ltd**

Preface

Biology is the story of life on earth. It is the science of life forms and living processes. Biological systems often appear to challenges and opportunities of physical laws that govern the behaviour of matter and energy in our world. Biology is offered as a subject at undergraduate and postgraduate levels in all the Indian Agricultural Universities. It is also offered at graduate level in biology and postgraduate level in botany in conventional universities. There are several books especially of foreign authors on the subject. However, the syllabi of Indian Universities are not covered by any single source. Moreover, books of foreign origin rarely cover Indian examples and their high price cannot afford by majority of Indian students. The prime objective of writing the present book "**BIOLOGY: FUNDAMENTAL PRINCIPLES**" is to help Indian students. The main features of this book are as follows:

1. This book covers in one compact volume the courses offered by various Indian Universities (Conventional & Agricultural) at graduate and postgraduate levels.
2. The language is simple and easily understandable.
3. The presentation is point wise and step by step with recent Indian examples.
4. The glossary has been presented at the end of each chapter which will help students in revision of respective chapter.
5. Efforts have been made to provide up-to-date information about various fundamental principles, methods and their practical implications especially in crop improvement.
6. Inclusion of five special chapters' *viz.,* Origin of life, Taxonomy and Systematics, Nomenclature, Introduction to Genetics and Plant systematic - Study of families, is a unique feature of this book. These

topics are of great significance in the modern context, but are lacking in almost all the previously published books on plant breeding.

Hope, this volume will be useful for the students, teachers and researchers in the field of Botany. It will also be useful for those appearing in competitive examinations. Suggestions for further improvement of this book shall be highly appreciated. Hopefully along with changes in the teaching and learning context, to be brought out in the next few years, the next edition of this book will reveal more integration of botany and truly reflect the true nature of biology - the future science of man by man and for man.

The cooperation extended by my family members during the completion of this project is highly appreciable. I am extremely thankful to all those who are help to me directly and indirectly for several constructive suggestions, encouragement and whole hearted support during completion of this book.

We also express our word of appreciation for the "***Shiv-Shambho Ganesh Pratisthan***" for their moral support.

Last but not the least, a million thanks to lord ***"GANESHA"***, who blessed me to do this task and made every job success. We express our sincere thanks to Publishers (Astral International Pvt Ltd.), to arrange early publication of this volume.

Authors

Contents

Chapter 1

Introduction to the Life or Living World

INTRODUCTION

Biology is the science of life forms and living processes. Biology (bios- living, logos- study) is a scientific branch that is related with the living creatures and their body's functions. There can be seen a huge variety of living creatures on this planet.

The distinction between Living Creatures and Non-Living matter was seen by early man. Definite arrangement of identification of organisms and further their nomenclature started later. This prompted the acknowledgment of sharing similarity among life forms. Man could perceive that all the present day living beings are identified with each other furthermore to those which ever lived on this planet. The parts of this unit manage the qualities of living creatures and their characterization. Living organisms show a great biodiversity and are classified into different kingdoms - Monera, Protista, Fungi, Plantae and Animalia.

How wonderful is the living world! The wide range of living types is amazing. The extraordinary habitats in which we find living organisms, be it cold mountains, deciduous forests, oceans, fresh water lakes, deserts or hot springs, leave us speechless. The beauty of a galloping horse, of the migrating birds, the valley of flowers or the attacking shark evokes awe and a deep sense of wonder. The ecological conflict and cooperation among members of a population and among populations of a community or even the molecular traffic inside a cell make us deeply reflect on - what indeed is life? This question has two implicit questions within it. The first is a technical one and seeks answer to what living is as opposed to the non-living and the second is a philosophical one and seeks answer to what the purpose of life is. As scientists, we shall not attempt answering the second question. We will try to reflect on - what is living?

The Living World

Several type of living beings are found in various sorts of living spaces like sea, freshwater bodies, backwoods, cool mountains, deserts, warm water springs and so on. This makes us think what is life? The answer requires separating the living and the non-living. When we try to define 'living', we conventionally look for distinctive characteristics exhibited by living organisms. Growth, reproduction, ability to sense environment and mount a suitable response come to our mind immediately as unique features of living organisms. One can add a few more features like metabolism, ability to self-replicate, self-organise, interact and emergence to this list. Let us try to understand each of these.

Characteristics of Life

It is extremely hard to characterize "living" and traditionally different qualities common to every single living being are to be recognized. The seven characteristics of life are recorded beneath:

1. **Cell unit of life:** Basic structural and functional unit of life is cell.
2. **Cellular Association:** Living things are composed of single cell (Unicellular Organisms) or many cells (Multi-cellular Organisms) which associate with each other to perform vital functions of the body.
3. **Response to Stimuli or Responsiveness to the environment:** All living things can respond and adapt themselves to any change in the environment.
4. **Growth and development:** Increase in mass and increase in number of individuals are twin characteristics of growth. All living things grow, develop and eventually die.
5. **Reproduction Ability:** reproduction refers to the production of progeny possessing features more or less similar to those of parents. Living things reproduce either sexually or asexually to produce off-springs of their own kind.

6. **Metabolism:** Living things need energy to carry out their vital life processes. All plants, animals, fungi and microbes exhibit metabolism. The sum total of all the chemical reactions occurring in our body is metabolism.

7. **Maintain homeostasis:** Living things are having ability to resists to change the buffering capacity.

8. **Passing traits onto offspring:** Every life or living in which transfer the characters from one generation (parents) to the next generation (offspring).

It is evaluated that more than 5 million species are available on earth. Of these around 1.7 million species are known and portrayed. Consistently a few new animal groups are portrayed and added to the rundown. Since investigation of all living beings is about incomprehensible, they are ordered into groups for the advantageous study.

BUILDING BLOCK OF LIFE AND THEIR FUNCTIONS

Living life form is shaped of many sorts of inorganic and also organic biomolecules. Inorganic compounds incorporate water & minerals and are constantly micro biomolecules (little measured, low atomic weight, promptly dissolvable in water and diffusible) while organic particles might be smaller scale (e.g. mono-sugars, amino acids) or macro biomolecules (expansive size, high sub-atomic weight, insoluble or somewhat dissolvable and non-diffusible e.g. fats, proteins & nucleic acids).

- ✓ **Function of Water:** Water frames 70-90% of the cellular pool. It frames 65% (around 66%) of human body. It is framed of H and O in the proportion of 2:1. 95% of water is found in Free State and 5% in consolidated shape in the cell. Water helps in supporting the life forms. Hence, water is known to be a remedy or support of lip as life is impractical without water.

- ✓ **Function of Oxygen:** Oxygen is for the most part used in vigorous cell breath of the supplements inside the mitochondria to create energy-rich ATP particles so is fundamental forever. Without oxygen, just 5% of energy accessible is discharged.

- ✓ **Function of salt** (Sodium chloride): Sodium chloride assumes a vital part in metabolic elements of body particularly when in ionic frame.

- ✓ **Function of Carbohydrates:** Carbohydrates are organic compounds shaped of C, H and O for the most part in the proportion of 1:2:1. These are normally called saccharides (Saccharin - sugar) Carbohydrates are the primary storage atoms and most life forms utilize starches as a vital fuel, breaking these securities and discharging energy to maintain life.
- ✓ **Function of Proteins:** Proteins are polymeric compounds framed by interlinking of amino acids (monomers) by peptide bonds. Out of around 100 sorts of amino acids, just 20 sorts of amino acids are of natural significance, so are called Magic 20. Proteins assume a key part in the arrangement of structures in living beings. Like starch and fat protein can be separated with the arrival of energy. Protein is not stored all things considered in the body and it is regularly just utilized as a significant wellspring of energy in states of starvation.
- ✓ **Function of lipids:** Lipids involve a noteworthy group of insoluble hydrocarbons having many capacities. These are polymers of alcohols (e.g. glycerol) and fatty acids interlinked by ester bonds. Complex lipids, for example, genuine fats are critical organic atoms that are utilized to give energy. Fats in creatures likewise give security from warmth loss.
- ✓ **Function of Nucleic Acid:** These are polymers of nucleotides interlinked by phosphodiester bonds, alleged polynucleotides. Every nucleotide is framed of 3 segments: a pentose sugar (e.g. deoxyribose in DNA and ribose in RNA), a phosphate group and an inorganic nitrogen base (a purine or a pyrimidine).

DIVERSITY IN THE LIVING WORLD

If we look around we will see a large variety of living organisms, be it potted plants, insects, birds, your pets or other animals and plants. There are also several organisms that you cannot see with your naked eye but they are all around you. If you were to increase the area that you make observations in, the range and variety of organisms that you see would increase. Obviously, if you were to visit a dense forest, you would probably see a much greater number and kinds of living organisms in it. Each different kind of plant, animal or organism that you see, represents a species. The number of species that are known and described range between 1.7-1.8 million. This refers to biodiversity or the number and types of organisms present on earth. We should remember here that as we explore new areas and even old ones, new organisms are continuously being identified.

'Biological diversity' means the variability among living organisms from all sources including, inter alia, terrestrial, marine and other aquatic ecosystems and the ecological complexes of which they are a part; this includes diversity within species, between species and of ecosystems.

GLOSSARY

Biology: It is a scientific branch that is related with the living creatures and their body's functions.

Cell: Basic structural and functional unit of life.

Reproduction: Refers to the production of progeny possessing features more or less similar to those of parents.

QUESTIONS

Q. 1. Explain in brief about the living world.

Q. 2. Explain in detail characteristics of life.

Q. 3. Enlist and explain in brief building blocks of life along with its functions.

Chapter 2

Origin of Life

INTRODUCTION

Primitive Earth was very different than the way things are now. There were probably many oceans and seas with many hot vents at the bottom of these waters and quite a bit of volcanic activity on land. The atmosphere most likely contained water, methane, ammonia and hydrogen, unlike our current atmosphere, which is mostly nitrogen and oxygen. Hence, in that condition, the origin of life on Earth is a highly curious thing. In fact, many scientists have dedicated their entire lives to finding out how life came to be on Earth. There are a few key experiments that we will look at in order to gain an understanding of how scientists have best hypothesized how life started on Earth, but let's first take a trip back in time - about four billion years ago.

THEORIES OF ORIGIN OF LIFE

Various theories have been put forth to explain the phenomenon of origin of life. A few of them were only speculations while others were based on scientific grounds. These theories are given below

- Theory of special creation
- Theory of spontaneous generation or Abiogenesis
- Biogenesis
- Cosmozoic theory
- Theory of sudden creation from inorganic material
- Naturalistic theory

OPARIN-HALDANE THEORY OF CHEMICAL ORIGIN OF LIFE

- Oparin (1924) proposed that, "life could have originated from non-living organic molecules."
- He believed in Biochemical origin of life. Haldane (1929) also stated similar views. Oparin greatly expanded his ideas and presented them as a book "The origin of life" in 1936.
- According to this theory, the Earth originated about 4,500 million years ago. When the earth was cooling down, it had a reduced atmosphere. In this primitive atmosphere nitrogen, hydrogen, ammonia, methane, carbon mono-oxide and water were present. Energy was available in the form of electric discharges by lightning and ultraviolet rays. As soon as the earth crust was formed, it was very much folded. Torrential rains poured over the earth for centuries and were deposited in deep places.
- Miller's Experiment: An American Biologist, Stanley Miller (1953) performed an experiment under supports Oparin's theory of origin of life. He believed that basic compounds which are essential for life can be synthesised in the laboratory by creation in the laboratory, on a small scale, the conditions which must have existed at the time of origin of life on earth.
- Miller took a flask and filled it with methane, ammonia and hydrogen in proportion of 2:1:2 respectively at 0°C. This proportion of gases probably existed in the environment at time of origin of life. This flask was connected with a smaller flask that was filled with water, with the help of glass tubes. In the bigger flask, two electrodes of tungsten were fitted. Then a current of 60,000 volts was passes, through gases containing bigger flask for seven days. At the end of seven days, when the vapours condensed, a red substance was found in the U-tube. When this red substance was analysed, it was found to contain amino acids, Glycine and nitrogenous bases which are found in the nucleus of a cell.
- The entire process of the origin of life, as proposed by Oparin, can be summarised as under,

The Chemical Evolution

- ❖ Step 1: Formation of simple molecules
- ❖ Step 2: Formation of Simple organic compounds
- ❖ Step 3: Formation of complex organic compounds
- ❖ Step 4: Formation of nucleic acids and nucleoproteins

The Organic Evolution

- ❖ Step 5: Formation of Coacervates
- ❖ Step 6: Formation of Primitive cell
- ❖ Step 7: Origin of autotrophism
- ❖ Step 8: Origin of Eukaryotic cells

EVIDENCES OF ORGANIC EVOLUTION

There are the different evidences in favour of organic evolution which are as given below

1. Evidences from Classification
2. Evidences from Comparative Anatomy
   i). Analogy and Homology
   ii). Vestigeal organs
3. Evidences from Physiology
4. Evidences from Serology
5. Evidences from Embryology
6. Evidences from Palaeontology
7. Evidences from geographic distribution
8. Evidences from Genetics

1. **Evidences from Classification:** All the known living animals and plants have been classified into various species, genera, families, order, classes, phyla and kingdoms. The classification of a particular animal is attempted only after its extensive study.
2. **Evidences from Comparative Anatomy:** In all the living animals, the basic substance of life is Protoplasm. If the species had been created separately, then there should be no relationship in the various organs and systems of animals. But on the contrary, we see that large number

of animals although unlike in appearance show most of the systems and organs made on the same plan. The resemblance is very close in the members of the same group.

3. **Evidences from Physiology:** Various types of chemical tests exhibit many basic similarities in physiological and chemical properties that show a physiological relationship among animals.
4. **Evidences from Serology:** This is a method by which the reactions of blood serum are observed. From the blood are also extracted the crystals of oxyhaemoglobin. The structure differs in different vertebrates, but in a definite order. The reaction is nearly identical in man and anthropoid monkeys, but slightly less identical with other mammals.
5. **Evidences from Embryology:** With the exception of a few, every multi-cellular animal originates from a zygote. The development from zygote to adult shows many similarities in various organisms. The development is termed as ontogeny
6. **Evidences from Palaeontology:** The study of fossils and their interpretation forms one of the great evidences of evolution. An Italian scientist, Leonardo da Vinci, was the first person to recognize their importance and said they were either remains of organisms of their impressions on some sort of clay or rock.
7. **Evidences from geographic distribution:** If the study of horizontal distribution of animals on the face of this earth is made, it would be seen that animals are not evenly distributed. Two identical places with the same climate and vegetation may not have same sort of animal fauna
8. **Evidences from Genetics:** Gregor Johann Mendel in 1866 published his work on experimental breeding. He bred two individuals differing in certain well defined characters and observed the ratio in which various contrasting parental characters appeared in successive generations.

THEORIES OF ORGANIC EVOLUTION

Lamarckism: Lamarck (1744 -1829) was one of the most brilliant stars on the horizon of the history of evolution. He was the first naturalist to put forward a general theory of evolution in his famous book. 'Philosophic Zoologique' published in 1809. His evolutionary theory may be summarised in the form of following laws:

(a) The internal forces of life tend to increase the size of an organism.

(b) The necessity in animals to produce new structures.

(c) The effect of use and disuse.

(d) Inheritance of acquired characters.

Darwinism: Charles Robert Darwin was undoubtedly the first naturalist who put the idea of organic evolution on sound footing. His statements and theories were based upon practical experiences and large number of proofs which he collected directly from the nature. His main ideas about the evolution are given below

(a) Over - production of offspring

(b) Limited supply of food and shelter

(c) Struggle for existence

  (i) Intra-specific

  (ii) Inter-specific

  (iii) Environment

(d) Survival of the fittest

(e) Universal occurrence of variations

(f) Inheritance

(g) Natural selection

GLOSSARY

Serology: The science that studies the reaction between antigens and antibodies in serum.

Embryology: The scientific study of embryos.

Palaeontology: Study of the forms of life existing in prehistoric or geologic times, especially as represented by fossils.

QUESTIONS

Q. 1. Enlist theories of origin of life. Explain the Oparin - Haldane theory of chemical origin of life.

Q. 2. Explain in detail the theories and evidences of organic evolution.

Chapter 3

Classification

INTRODUCTION

Classification is a process related to categorization, the process in which ideas and objects are recognized, differentiated and understood. A classification is an ordered set of related categories used to group data according to its similarities. It consists of codes and descriptors and allows survey responses to be put into meaningful categories in order to produce useful data. A classification is a useful tool for anyone developing statistical surveys. It is a framework which both simplifies the topic being studied and makes it easy to categorise all data or responses received.

OBJECTIVES OF CLASSIFICATION

- To identify and describe all the possible types of species.
- To arrange the species in various categories on the basis of their similarities and dissimilarities.
- To evolve a truly natural or phylogenetic system this should indicate origin and evolution of the species.
- To help for easy identification of organisms.

HISTORY OF CLASSIFICATION

1. Aristotle: Father of biology & zoology.

2. Theophrastus:

- He is known as father of ancient plant taxonomy and botany.
- Theophrastus wrote many books on plants. Few of them are as given below
- Historia plantarum

- Causes of plants
- Enquiry into plants
- Theophrastus gave names and descriptions of 480 plants in his book "Historia plantarum".
- Theophrastus proposes the first classification of plant kingdom. He classified plant kingdom in to four groups on the basis of growth habit. (a) Trees (b) Shrubs (c) Under shrubs (d) Herbs

3. Carolus Linnaeus: [1707 - 1778]

- He is known as father of taxonomy, father of plant and animal taxonomy.
- Linnaeus gave the two kingdom system classification. He grouped plants and animals into kingdom plantae and kingdom Animalia respectively.
- Linnaeus wrote many books. Some important books are given below
  - Hortus uplandicus - First book
  - Flora lapponica
  - Philosophia botanica
  - Critica botanica
  - Systema naturae (1737)
  - Genera plantarum
  - Species plantarum -last book (1753)

4. A.P. De Candolle:

- He wrote the book – 'Theories Elementaire de la botanique'.
- He was the first to propose the significance of vascular tissue in taxonomy. On this basis of vascular tissue he classified plants into two groups.
- Cellular plants (Non-vascular plants) - This group includes Thallophyta and Bryophyta
- Vascular plants - This group includes Pteridophyta, Gymnosperm and Angiosperms.

5. George Bentham (1800 -1884) and Joseph Dalton Hooker: (1817 -1911)

- ❖ Both Bentham and Hooker were related to Royal botanical garden.
- ❖ They wrote the book 'Genera plantarum' (1862 - 1883).
- ❖ In this book, Bentham and Hooker gave the biggest and natural classification of spermatophyta *i.e.* plants with seeds
- ❖ In Genera plantarum, there is description of 202 families. In it, basically the description of plants with seeds is present.

6. A. W. Eichler:

- ❖ 'Syllabus de vorlesungen uber phanerogamen kunde' - Book written by Eichler.
- ❖ In this book, Eichler gave the first phylogenetic classification of plant kingdom.
- ❖ The classification of Eichler is very little phylogenetic.

In this way Eichler classified plant kingdom into five divisions and arranged them in the order of evolution (Phylogeny).

7. Engler (1844 - 1930) & Pranti (1849 - 1893):

- ❖ Book – 'Die Naturlichen Pflanzen Familien'.
- ❖ He gave the phylogenetic classification of plant kingdom. This classification was more phylogenetic as compared to Eichler's classification.

8. Oswald Tippo:

- ❖ Proposed the biggest phylogenetic classification of plant kingdom.
- ❖ This classification is the complete classification of plant kingdom.
- ❖ This is the most acceptable classification for books and study.
    - Cyanophyta: Blue green alage
    - Euglenophyta: Euglenoids
    - Chlorophyta: Green algae
    - Chrysophyta: Yellow-green algae
    - Pyrrophyta: Dinoflagellates & Diatoms
    - Phaeophyta: Brown algae
    - Rhodophyta: Red algae

- Schizomycophyta: Bacteria
- Myxomycophyta: Slime molds (False fungi)
- Eumycophyta: True fungi

9. Karl Menz:

- He showed the importance of serology in taxonomy.
- Similarities and dissimilarities in structure of proteins help to know the phylogenetic relationship of living beings. Living organisms which are phylogenetically close relatives have more similarities in their proteins.
- Organisms which are distantly related have different proteins.

10. Haeckel:

- Haeckel gave the three kingdom (Protista, Plantae, Animalia) system of classification.
- He established the kingdom Protista.
- He grouped those living organisms in Protista which did not have tissues.
- Kingdom Protista: Prokaryotes, Protozoa, porifera, Algae & fungi.

BIOLOGICAL CLASSIFICATION

Biological Classification is the scientific study of arranging organisms into group and subgroup on the basis of their similarities and dissimilarities and placing the group in a hierarchy of categories.

The purpose of biological classification is to organise the vast number of known plants into categories that could be named, remembered and studied.

TYPES OF BIOLOGICAL CLASSIFICATION

There are three main types of biological classification - artificial system of classification, natural system of classification and phylogenetic system of classification.

1. Artificial System of Classification

It is a system of classification which uses one or two morphological character for grouping of organisms. Some artificial systems have used habot and habitat for this purpose. Aristotle (350 BC) divided animals into two categories, enaima (with red blood) and anaima (without red blood).

Aristotle also classified animals on the basis of their habitat- aquatic (e.g, fish and whale), terrestrial (e.g, reptiles, cattle) and aerial (e.g. birds, bat). Pliny the Elder (23-79 A.D.) used artificial system of classification for both plants and animals dividing them into land, air and water. Pliny distinguished animals into flight band non-flight ones. Flight animals included bats, birds and insects.

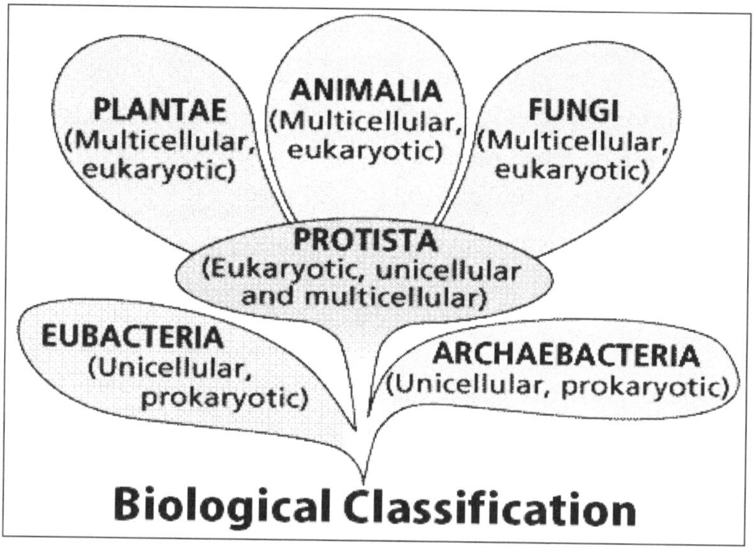

Fig. 2.1. Biological Classification

2. Natural System of Classification

It is a system of classification which takes into consideration comparable study of a number of characters so as to bring out nature similarities and dissimilarities and hence nature relationship among the organisms. The system employs those characters which are relatively constant. They include morphological characters, anatomical characters, cytological characters, physiology, ontogeny or development, reproduction, cytochemistry and biochemistry, experimental taxonomy etc. the characteristics are helpful in bringing out maximum number of similarities in a group and comparable differences with other group of organisms. For example, mammals are characterised by the presence of mammae, birds possess wings, feathers, pneumatic bones, ovipary, 4- chambered. They are coldblooded.

3. Phylogenetic System of Classification

Classification based on evolutionary relationship of organisms is called phylogenetic system of classification. It is based on the evolutionary concept from Darwin's book, on the origin of species by means of natural selection. The preservation of favoured races in the struggle for life. It reflects the true relationships among the organisms. First phylogenetic system was proposed by Engler and Prantl (1887-99). Zoologists believe that since similarity in structure represents close evolutionary relationship, their natural classification represents evolutionary and phylogenetic classification.

Five Kingdom Classification (from 1969 to 1990)

In order to develop phylogenetic classification, R.H. Whittaker (1969), an American taxonomist, divided all the organisms into five kingdoms. As the viruses are on the border line of living and non-living, they have been left out. Whittaker has used five criteria for delimiting the different kingdoms.

- Complexity of cell structure, prokaryotic and eukaryotic.
- Complexity of body structure or structural organization, unicellular and multi-cellular.
- Mode of nutrition which is divergent in multi-cellular kingdoms - photoautotrophy in plantae, absorptive heterotrophy in fungi and ingestive heterotropy in animalia. Photoautotrophy totrophic nutrition is also called halophytic nutrition while ingestive heterotrophy is known as holozoic nutrition. Absorptive heterotropy is saprobiotic (saprophytic) nutrition.
- Ecological life style like producers (plantae), decomposers (fungi) and consumers (animalia).
- Phylogenetic relationships.

Three Domains of Life or Six Kingdom Classification (1990)

- The three - domain system is a biological classification which was introduced by Carl Woese, a professor in the department of microbiology, university of Illinois, Urbana- Champaign in 1990 that divides cellular life forms into archaea, bacteria and eukarya domains.
- It emphasizes the separation of prokaryotes into two groups, originally called eubacteria (now bacteria) and archaebacteria (now archaea) because of their fundamental differences, Woese argued that each of the two arose separately from an ancestor with poorly developed genetic machinery, often called a progenote.

Table 2.1. Characteristics of Whittaker's five kingdom system of classification

Sr. No.	Characters	Monera	Protista	Fungi	Plantae	Animalia
1	Cell type	Prokaryotic	Eukaryotic	Eukaryotic	Eukaryotic	Eukaryotic
2	Cell wall	Non-cellulosic polysaccharide+ amino acid)	Present in some (various types)	Present (non- cellulosic)	Present (cellulose)	Absent
3	Chloroplast	Absent	Present in some	Absent	Present	Absent
4	Mitochondria	Absent	Present in some	Present	Present	Present
5	Nuclear membrane	Absent	Present	Present	Present	Present
6	Tissue or multicellularity	Absent	Absent	Present but limited	Present in all forms	Present in all forms
7	Motility	Bacterial flagella, gliding or non-motile	Cilia, flagella amoeboid or contractile fibrils	Cilia, flagella in some, none in most of the forms	Cilia and flagella in lower forms, absent in most of the forms	Cilia and flagella, contractile fibrils
8	Mode of nutrition	Autotrophic-chemosynthetic and photo- synthetic, heterotrophic	Photosynthesis and heterotrophic	Heterotrophic, saprophytic and parasitic absorptive	Autotrophic by photosynthes	Heterotrophic by ingestion
9	Reproduction/ means of genetic recombination	Conjugation transduction transformation or none	Syngamy and meiosis, conjugation or none	Fertilization and meiosis, dikaryosis or none	Fertilization and meiosis	Fertilization and meiosis
10	Nervous system	Absent	Primitive for conduction stimuli	Absent	Absent	Present, often complex

- In fact the three - domain system is loosely based on the traditional five- kingdom system but divides the monera into two domains, leaving the remaining eukaryotic kingdoms in the third domain.
- It is actually a six kingdom classification.

1. Archaea domain

The domain contains prokaryotic organisms which have a monolayer core of lipids in the cell memebrane and distinct nucleotides in their 16S RNA. It contains a single kingdom.

Kingdom archaebacteria

The kingdom contains early prokaryotes which live in extreme environments.

For Example:

(a) Methanogens - metabolize hydrogen and carbon dioxide into methane.

(b) Halophiles - live in salt.

(c) Thermoacidophiles - live in acid high temperatures (upto 110^0 Celsius).

2. Bacteria domain

The domain contains prokaryotes which lack membrane covered cell organelles but do have a sort of micro chambers for separating various activities. There is a single kingdom.

Kingdom eubacteria

The domain contains diverse type of bacteria having peptidoglycan cell wall, glycogen as food reserve, naked DNA coiled to form nucleoid, absence of sap vacuoles and presence of 70S ribosomes. Some common group are bacteria, mycoplasma, actinomycetes, rickettsiae, spirochaetes, firmicutes and cyanobacteria.

3. Eukarya domain

The domain contains eukaryotic organisms which originated by endosymbiotic association between some archaebacteria and eubacteria. It has four kingdoms which are given below

Kingdom protista

Protista are eukaryotic and can be unicellular or simple multi-cellular. They reproduce sexually or asexually. Important examples of protists include the organism known as *Plasmodium* (which causes malaria), *Amoeba* and *Euglena*. There are two major groups of protists which include the Protozoans, whose cells are similar to animal cells in that they do not have cell walls and the plant-like cells which do have cell walls and are similar to algae.

Kingdom fungi

Fungi are eukaryotic organisms that can be multi-cellular or unicellular. Mushrooms and moulds are examples of multi-cellular fungi and yeast is an example of a unicellular fungi. All fungi have a cell wall made of chitin. They are non-motile (not capable of movement) and consist of threads called hyphae. Fungi are a heterotrophic organism which means they require organic compounds of carbon and nitrogen for nourishment. They are important as decomposers (saprophytes) and can be parasitic. They store carbon as glycogen, not in the form of starch. Fungi reproduce sexually and asexually by spore formation. An important example of a useful fungus is *Penicillium* (a fungus which was used to make penicillin, one of the most powerful antibiotics ever created).

Kingdom plantae

Organisms belonging to the plant kingdom are eukaryotic and multi-cellular organisms. They have a distinct cell wall made of cellulose. Cells are organised into true plant tissues. Plants contain plastids and photosynthetic pigments such as chlorophyll. They are non-motile. Plants make their own food by photosynthesis and are therefore said to be autotrophic. Plants undergo both sexual and asexual reproduction. They store food as starch. Important examples of plants are mosses, ferns, conifers and flowering plants.

Kingdom animalia

Members of the animal kingdom are eukaryotic and multi-cellular but have no cell wall or photosynthetic pigments. They are mostly motile and they are heterotrophic, which means they must feed on other organisms and cannot make their own food. They reproduce sexually or asexually. Animals store carbon as glycogen and fat. Important examples of this kingdom include: Porifera (Sponges), Cnidaria (Jellyfish), Nematoda (nematode worms), Platyhelminthes (Flatworms), Annelidas (segmented worms), Mollusca (Snails and Squid), Echinodermata (Starfish), Arthropoda (Insects and Crustaceans), Chordata (includes all the vertebrates: fish, amphibians, reptiles, birds, mammals).

Table 2.2. Resent Classification System.

Linnaeus 1735	Haeckel 1866	Chatton 1925	Copeland 1938	Whittaker 1969	Woese et. al. 1990	Cava-lier-Smith 1998	Cava-lier-Smith 2015
2 kingdoms	3 kingdoms	2 empires	4 kingdoms	5 kingdoms	3 domains	2 empires, 6 kingdoms	2 empires, 7 kingdoms
(not treat-ed)	Protista	Prokaryota	Monera	Monera	Bacteria	Bacteria	Bacteria
					Archaea		Archaea
		Eukaryota	Protoctista	Protista	Eucarya	Protozoa	Protozoa
						Chromista	Chromista
Vegeta-bilia	Plantae		Plantae	Plantae		Plantae	Plantae
				Fungi		Fungi	Fungi
Animalia	Animalia		Animalia	Animalia		Animalia	Animalia

GLOSSARY

Biological Classification: It is the scientific study of arranging organisms into group and subgroup on the basis of their similarities and dissimilarities and placing the group in a hierarchy of categories.

QUESTIONS

Q. 1. Give the objectives of the classification and explain in brief history of classification.

Q. 2. Explain in detail types of biological classification.

Q. 3. What are the characteristics of Whittaker's five kingdom classification system?

Chapter 4

Taxonomy and Systematics

TAXONOMY

People who live close to nature usually have an excellent working knowledge of the elements of the local fauna and flora important to them and also often recognize many of the larger groups of living things (e.g., fishes, birds and mammals). Their knowledge, however, is according to need, and such people generalize only rarely. In a broad sense it is the science of classification, but more strictly the classification of living and extinct organisms - *i.e.,* biological classification. The term is derived from the Greek *taxis* means arrangement and *nomos* means law. Hence, Taxonomy is the methodology and principles of systematic botany and zoology and sets up arrangements of the kinds of plants and animals in hierarchies of superior and subordinate groups.

DEFINITION

"Taxonomy is the study of principles and procedures of classification."

Taxonomy includes study of following 4 points:

1. Identification: Identification of living organisms. Nomenclature or naming is only possible when the organism is described correctly and we know to what organism the name is attached to.
2. Nomenclature: Nomenclature of living organisms. There is a need to standardise the naming of living organisms such that a particular organism is known by the same name all over the world.
3. Classification: Classification of living organisms in groups. Classification is the process by which anything is grouped into convenient categories based on some easily observable characters.
4. Affinities: Study of inter relationship between living organisms.

TYPES OF TAXONOMY

1. **Alpha taxonomy or classical taxonomy:** It is based on external morphology, origin and evolution of plants.
2. **Beta taxonomy or Explorative taxonomy:** Besides external morphology, it also includes internal characters' like embryological, cytological, anatomical characters etc.
3. **Omega taxonomy or encyclopaedic taxonomy:** Omega taxonomy has widest scope. It is based on all the information or data available about plants.
4. **Cytotaxonomy:** The use of cytological characters of plants in classification or in solving taxonomic problems is called cytotaxonomy. Cytological characters constitute an important aid to plant taxonomy, especially in determining affinities at the generic and infra-generic levels.
5. **Chemotaxonomy:** The use of chemical characters of plants in classification or in solving taxonomic problems is called chemotaxonomy or chemical taxonomy. It is based on the chemical constitution of plants. The fragrance and taste vary from species to species. The basic chemical compounds used in chemotaxonomy are alkaloids, carotenoids, tannins, polysaccharide, nucleic acids, fatty acids, amino acids, aromatic compounds etc.
6. **Karyotaxonomy:** Based on characters of nucleus and chromosomes. Pattern of chromosomal bands (dark bands and light bands) is most specific characters.

Table. 3.1. Differences between Classical Taxonomy and Modern Taxonomy

Sr. No.	Classical Taxonomy	Modern taxonomy/ Neo systematic
1.	It deals with morphological species.	It deals with biological species.
2.	Species is considered to be static.	Species is considered to be dynamic.
3.	It has a typological concept.	It has a population or biosystematics concept.
4.	It doesn't study evolution and inter-relationship of species.	It studies primitiveness, advancement and inter-relationships of species

TAXONOMIC CATEGORIES

Classification is not a single step process but involves hierarchy of steps in which each step represents a rank or category. Since the category is a part of overall taxonomic arrangement, it is called the taxonomic category and all categories together constitute the taxonomic hierarchy.

It is also called Linnaean hierarchy or taxonomic hierarchy or Taxonomic classification. It was first proposed by Linnaeus. Hierarchy of categories is the classification of organism in a definite sequence of categories (taxonomic categories) in a descending order starting from kingdom. The number of similar characters of categories decreases from lowest rank to highest rank. The hierarchy includes seven obligate categories - kingdom, division or phylum, class, order, family, genus and species. The categories are arranged in descending sequence keeping the kingdom at the top. In order to make taxonomic position of species more precise, certain categories have been added to this list. They are called intermediate categories.

There are 7 main taxonomic categories. They are obligate categories *i.e.,* they are strictly used at the time of any plant classification.

1. **Species:** Species (used both as singular and plural) is a natural population of individuals or group of population which resemble one another in all essential morphological and also reproductive characters so that they are able to interbreed freely and produce fertile offspring. Mango is species *indica* of genus *mangifera* (*Mangifera indica*). Potato is species *tuberosum* of genus *solanum* (*Solanum tuberosum*).

2. **Genus:** It is a group or assemblage of relate species which resemble one another in certain correlated characters. Correlated Characters are those similar or common features which are used in delimitation of a taxon above the rank of species. All the species of genus are presumed to have evolved from a common ancestor.

3. **Family:** It is taxonomic category which contains one or more related genera. All the genera of a family have some common features or correlated characters. They are separable from genera of a related family by important and characteristic differences in both vegetative and reproductive features. Thus the genera of cats (felis) and leopard (panther) are included in the family felidae.

4. **Order:** The category includes one or more related families. Thus the family *solanaceae* is placed in the order polemoniales along with four other related families (*convolvulaceae, boraginaceae, hydrophyllaceae* and *polemoniaceae*). Similarly, the families' *fekidae* and *canidae* are included under the order carnivore along with *hyaenidae* (hyaenas) and *ursidae* (bears).

5. **Class:** A class is made of one or more relates order. For example, the class dicotyledoneae (dicotyledonae, dictoyledons) of flowering plants contains all dicots which are grouped into several orders (e,g. *Rosales, passiflorales, polemoniales, sapindales, ranales*, etc.) likewise, class mammalian of animals includes all mammals which range from bats (order *Chiroptera*), kangaroos (order *Marsupialia*). Rodents (order *Rodentia*), whales (order *Cetacean*), carnivores (order *Cornivora*) to great apes and man (order *Primate*).
6. **Division or phylum:** It is a category higher than that of class. The term phylum is used for animals while division is commonly employed for plants. A division or phylum is formed of one or more classes. The phylum chordate of animals contains not class mammalian but also aves (birds), reptilian (reptiles), amphibian (amphibians), cyclostomata, chondrichthyes, osteichthyes (fishes) etc.
7. **Kingdom:** It is the highest taxonomic category. All plants are included in kingdom plantae while all animals belong to kingdom animalia.

There are some extra categories like sub division, sub order, sub family, tribe, sub tribe etc. They are not regularly used. They are used only when they are needed.

CONCEPT OF SPECIES

Species is a smallest taxonomic category. It is basic unit of classification. John Ray (1627-1705), English naturalist, proposed the term and concept of species. To explain the species, different concepts were proposed, which are as follows:

(A) Biological concept of species

- Mayer proposed the biological concept of species.
- Mayer defined the 'species' in the form of biological concept.
- According to Mayer 'All the members that can interbreed. Among themselves and can produce fertile off springs are the members of same species'.
- But this definition of Mayer was incomplete because this definition is applicable to sexually reproducing living beings because there are many organisms that have only asexual mode of reproduction.

Example: Bacteria, *Mycoplasma* & *BGA*.

- The main character in determination of any species is interbreeding. But this, character is not used in taxonomy. In taxonomy, the determination of species is based on other characters. Example: Mainly morphological characters.
- In higher plants, the determination of species is mainly based on the morphology of flower (floral morphology). Because floral (reproductive) characters are more conservative as compared to vegetative (Root,Stem, Leaf) characters i.e. they do not shows any major changes.
- When the species is determined on the basis of interbreeding then it is called as biological species. Example: All the humans in this world can interbreed among themselves. So all the humans are the members of one biological species.
- When the determination of species is based on other characters then it is called as taxonomic species. Example: These 3 have same morphological characters. Therefore they belong to same taxonomic species i.e. one taxonomic species. But these three cannot interbreed among themselves. Therefore on the basis of interbreeding these are three biological species.

(B) Static concept of species

- The static concept of species was proposed by Linnaeus. According to Linnaeus species is unchangeable *i.e.* there is no change in the character of species. The species of present day are same as they were in past and they will remains same in future.
- Linnaeus believed in the 'Theory of Special creation'.
- Father Suarez - gave the principle of special creation. According to this theory, all the living organisms are created by God (Every life is created by God) and God gave the basic size and shape of all living organisms, they are still present in their actual former form. But Lamarck rejected this hypothesis.

(C) Dynamic concept of species

- It was proposed by Lamarck.

- According to this concept 'Species is always changeable'. Changes always occur in the characters of species from one generation to next generation. And these changes are known as evolution.

(D) Typological concept

- It was proposed by Aristotle and Plato.
- According to this concept 'There is a definite type or pattern of characters in the each species of every living organism and all the members of species show maximum resemblance with this pattern. (Typological concept is based on single individual of species) The species in which a fixed pattern of characters is present are called as monotypic species. Example: Bacteria, blue green algae.
- In many species, more than one type or pattern of characters are present. These are called Polytypic species or Macrospecies. Example: *Brassica oleracea* → Cauliflower, Cabbage, Knolkhol. Polytypic Species are of three types
    - **Biotype:** Members of same species inhabiting similar environment and having some genetic variations are known as biotypes. Variations found in these members are permanent. These members cannot interbreed among themselves. Example: Cauliflower, Cabbage, Knol-Khol are three biotypes of one species
    - **Eeotypes:** Members of same species inhabiting different environment and having some genetic variations are known as ecotypes. Variations are permanent. These members can interbreed among themselves but due to geographical barrier they cannot interbreed. Example: Crow (*Corvus splendense*) found in different regions are ecotype of one species
        - ✓ *Corvus splendense splendense* - Indian crow
        - ✓ *Corvus splendense insolence* - Myanmar crow
        - ✓ *Corvus splendense protegatus* - Srilankan crow
- **Eeads or Ecophenes:** Members of same species having some non-genetic variations due to environment is called Ecads. These variations are temporary. Example: Every living being. This hypothesis is believed to be most acceptable.

Definition related to species

- Linneon species: Those taxonomic species whose determination is based on morphology. They are called as linneon species. They are also called morpho-species or taxonomic species.

  Example: Most of species in taxonomy are linneon species

- Microspecies or Joardan's species: Those species in which variations are very less are called Joardan's species. They reproduce asexually so they have very less variations.
- Sibling species or Cryptic species: Members of species which are morphologically similar but reproductively isolated are known as Sibling species *i.e.* they cannot interbreed among themselves. Sibling species is one taxonomic species (because these members have similar morphology) but they are different biological species. Example: *Brassica oleracea* Allopatric species: Those species that are found in different geographical regions and have geographical barriers between them are known as allopatric species.
- Geographical barriers are hills, oceans, Himalayan Mountains.
- Sympatric species: The species found in similar geographical regions are sympatric species.
- Allochronk species: The species found in different time periods.

  Example: Man and Dinosaurs.

- Synchronic species: Those species that are found in same era.

  Example: Dinosaurs and Archaeopteryx.

- Palaeo species: Those species that are extinct now and are found in the form of fossils.

  Example: Dinosaurs

2. SYSTEMATICS (Branch Related with Taxonomy)

DEFINITION

"Systematics is a branch of Biology that deals with cataloguing plants, animals and other organisms into categories that can be named, remembered, compared and studied."

Study of only one organism of a group provides sufficient information about the remaining members of that group. Scientists connected with the study of systematics are called systematists or taxonomists.

Simpson, (1961) has defined systematics as the branch of biology that deals with the diversity of organisms at every level of classification.

BASICS OF SYSTEMATIC STUDY

- Characterization: The organism to be studied is described for all its morphological and other characteristics.
- Identification: Based on the studied characteristics, the identification of organism is carried out to know whether it is similar to any of the known group or taxa.
- Classification: The organism is now classified on the basis of its resemblance to different taxa. It is possible that the organism may not resemble any known taxa or groups. A new group or taxon is raised to accommodate it.
- Nomenclature: After placing the organism in various taxa, its correct name is determined. If the organism is new to systematics, it is given a new name based on rules and conventions of nomenclature.

GLOSSARY

Taxonomy: It is the study of principles and procedures of classification

Karyotaxonomy: Based on characters of nucleus and chromosomes.

Cytotaxonomy: The use of cytological characters of plants in classification or in solving taxonomic problems.

Chemotaxonomy: The use of chemical characters of plants in classification or in solving taxonomic problems is called chemotaxonomy.

Genus: It is a group or assemblage of relate species which resemble one another in certain correlated characters.

Family: It is taxonomic category which contains one or more related genera.

Order: The category includes one or more related families.

Class: A class is made of one or more relates order.

QUESTIONS

Q. 1. Define taxonomy. Give its types and write down difference between classical and modern taxonomy.

Q. 2. What do you mean by taxonomy? Explain in detail taxonomical catagories.

Q. 3. Define systematic. What are the basics of systematic?

Q. 4. Write short notes on concept of species.

Chapter 5

Nomenclature

INTRODUCTION

Frequently, the living beings around us are identified by local names which vary from spot to spot and dialect to dialect. Consequently there is a need to institutionalize naming them. Researchers have figured certain strategies to allot scientific name to every creature. **International Code for Botanical Nomenclature** (ICBN) and **International code for Zoological Nomenclature** (ICZN) were advanced to appoint scientific names for animals and plants separately. The investigative name has two segments generic name and the particular sobriquet. This arrangement of naming is called binomial nomenclature. It was given via Carolus Linnaeus and is utilized by scientists everywhere throughout the world. Examples: *Mangifera indica* (Mango); *Homo sapiens* (Man). In this case, *Mangifera* and *Homo* are the generic names; *indica* and *sapiens* are specific epithets.

1. Polynomial system

According to this system name of any plant consists of many words.

For Example: Caryophyllum

2. Binomial system

Binomial system was first proposed by Gaspard Bauhin in his book - 'Pinax Theatre Botanica'. The nomenclature is done by principle of priority. If two names are proposed for any plant after the 1753, the valid name is the earlier name proposed just after 1 May, 1753.

Main Rules of ICBN (International Code of Botanical Nomenclature)

Different rules which are given by International Code of Botanical Nomenclature (ICBN). These are follows:

1. According to binomial system name of any species consists of two names

   (i) Generic name - Name of genus

   (ii) Specific name - Trival name

   Example: Specific Name: Mango; Generic Name: *Mangifera indica*

2. In plant nomenclature (ICBN), tautonyms are not valid i.e. generic name and specific name should not be same in plants. Example: *Mangifera mangifera*

- But tautonyms are valid for animal nomenclature (ICZN-International Code of Zoological Nomenclature)

  Example: *Naja naja* (Indian cobra), *Raitus rattus* (Rat)

3. Length of generic mime or specific name should not be less than 3 letters and not more than 12 letters.

   Example: *Mangifera indica*

4. First letter of generic name should be in capital letter and first letter of specific name should be in small letter. Example: *Mangifera indica*

- But if specific name is based on the name of some person, its first letter should be in capital letter ego lsoetes Pantii

5. When written with free hand or typed, then generic name and specific name should be separately underlined. But during printing, name should be italized.

6. Name of scientist (who proposed nomenclature) should be written in short after the specific name Example: *Mangifera indica* Lin.

7. Name of scientist should be neither underlined nor written in italics, but written in roman, letters (simple alphabets)

8. If any scientist has proposed wrong name then his name should be written in bracket and the scientist who corrected the name should be written after the bracket.

   Example: *Tsuga canadensis* (Lin.) Salisbury

   Note: Linnaeus named this plant as *Pinus canadensis*

9. Scientific names should be derived from Latin or Greek languages because they are dead languages.

10. Type specimen (Herbarium Sheet) of newly discovered plant should be placed in herbarium (Dry garden).

11. Standard size of herbarium sheet is 11.5 × 16.5 inches

3. Trinomial System:

Proposed by Huxley and Stricklandt. According to this system name of any plant or species is composed of three names

(i) Generic name

(ii) Specific name

(iii) Sub-specifie name (Name of variety)

When members of any species have large variations then trinomial system is used. On the basis of dissimilarities this species is classified into subspecies

Example: *Brassica oleracea* var. *botrytis* (Cauliflower)

Brassica oleracea var. *capitata* (Cabbage)

Brassica oleracea var. *caulorapa* (Knol-Khol)

GLOSSARY

ICBN: International Code of Botanical Nomenclature

ICNB: International Code of Nomenclature for Bacteria

ICVN: International Code of Viral Nomenclature

ICNCP: International Code of Nomenclature for Cultivated Plants

QUESTIONS

Q. 1. Define nomenclature. What are the types of nomenclature?

Q. 2. What do you mean by binomial nomenclature? Give the different rules of binomial nomenclature according to ICBN.

Chapter 6

Cell

INTRODUCTION

The Cell (Latin *cella,* meaning small room) is the basic structural, functional and biological unit of all known living organisms. A cell is the smallest unit of life that can replicate independently and cells are often called the "Building Blocks of Life". The study of cells is called cell biology. Cell is the fundamental structural and functional unit of all living organisms. Anton Von Leeuwenhoek first saw and described a live cell. Robert Brown discovered the cork cell. Robert Brown later discovered the nucleus. Every living organism such as plants and animals are made up of cells. So, cell is considered as the basic unit of life.

TYPES OF CELL

1. Prokaryotes

Bacteria, blue green algae, mycoplasma and PPLO (Pleuro Pneumonia like Organisms) represents prokaryotes. The cell of prokaryotes is known as prokaryotic Cells. These cells are small in size and multiply at a faster rate. Prokaryotes have outermost covering known as cell wall, except in mycoplasma.

2. Eukaryotes

It includes protists, plants, fungi and animals. They have cell organelles separated by a membrane. They have well developed nucleus separated by a nuclear membrane.

CELL ORGANELLES

There are various cell organelles present in the cell. These are as given below

Cell wall

Many types of prokaryotic and eukaryotic cells have a cell wall. The cell wall acts to protect the cell mechanically and chemically from its environment, and is an additional layer of protection to the cell membrane. Different types of cell have cell walls made up of different materials; plant cell walls are primarily made up of cellulose, fungi cell walls are made up of chitin and bacteria cell walls are made up of peptidoglycan. Plant cell is surrounded by cell wall. Cell wall is made up of polysaccharide known as Cellulose. Cellulose provides strength and rigidity to the cell.

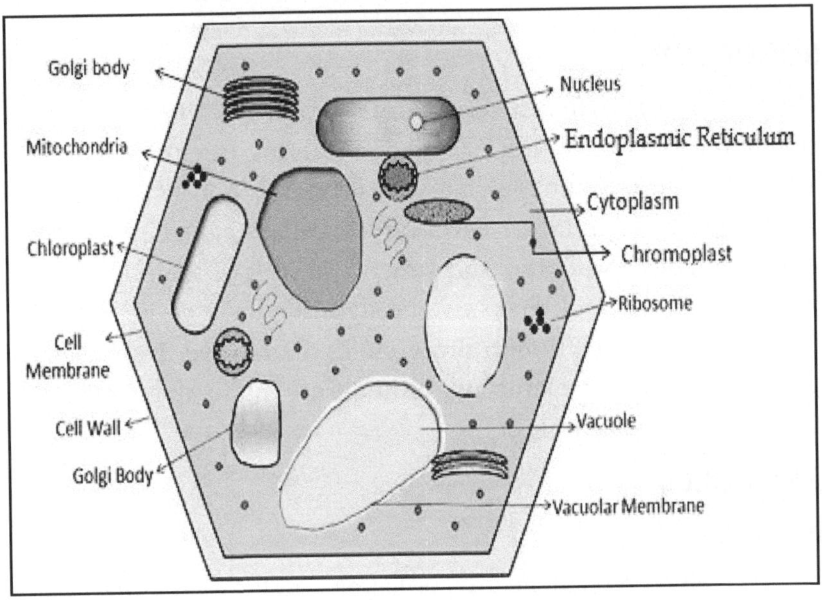

Fig. 6.1. Structure of Plant Cell

Plasma Membrane

In eukaryotes (higher organisms), cell is surrounded by a membrane known as plasma membrane or cell membrane. Plasma membrane is made up of lipids and protein. This membrane is selective or semi-permeable in nature, that is, it allows only certain molecules to enter the cell while remaining left outside the cell.

There are certain modifications of plasma membrane such as:

Bacteria have outermost envelope known as glycocalyx. Glycocalyx is followed by cell wall and then plasma membrane. It is protective in nature.

According to cell envelope or staining procedure, bacteria are classified as gram negative and gram positive. Bacteria that can be stained using gram stain are known as gram positive bacteria whereas those which are not stained are considered gram negative bacteria.

If glycocalyx is a loose sheath, then it is known as slime layer whereas if it is thick and tough it is known as capsule.

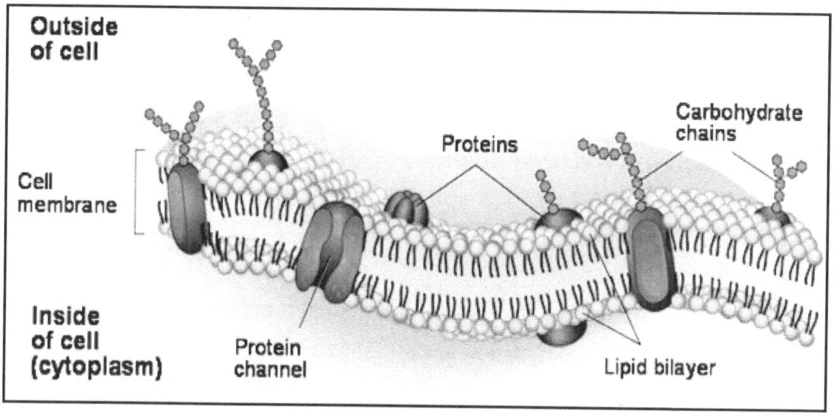

Fig. 6.2. Structure of Plasma Membrane

Ribosomes

These are the sites of protein synthesis. They are composed of two subunits- 50S and 30S subunit. These subunits together form 70S unit. 70S ribosome is present in bacteria. Eukaryotes contain 80S ribosome. The subunit of eukaryotic ribosomes is 60S and 40S. In prokaryotes, reserve material is stored in cytoplasm is the form of inclusion bodies.

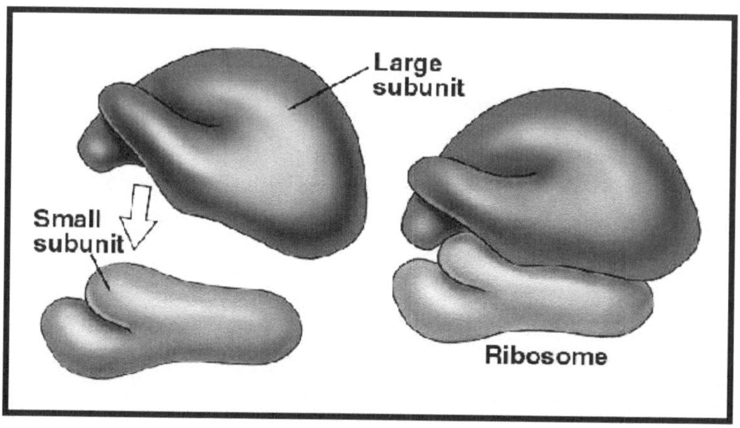

Fig. 6.3. Ribosome

Endomembrane System

It includes nucleus, endoplasmic reticulum, mitochondria, lysosomes, golgi apparatus, plastids in plants and vacuole.

Endoplasmic Reticulum (ER)

A network of tubules spread in a cytoplasm is known as endoplasmic reticulum or ER. Inside of the ER is known as lumen whereas outside the lumen is known as extra luminal. There are two types of ER- Rough ER and Smooth ER. Rough ER is covered by ribosome whereas smooth ER does not contain ribosome. Rough ER is involved in protein synthesis and smooth ER is involved in fatty acid synthesis and detoxification.

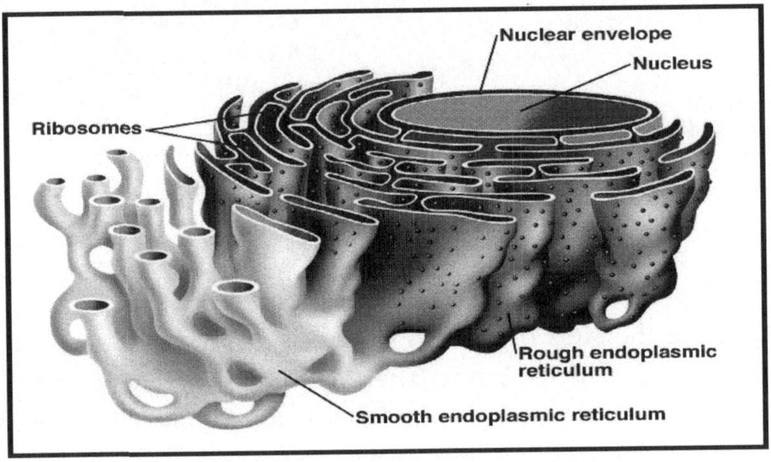

Fig. 6.4. 3D Structure of ER

Golgi apparatus

Camillo Golgi first observed this structure. They consist of many flat, disc-shaped sacs or cisternae of 0.5µm to 1.0µm diameter. They are arranged near the nucleus. The side facing the nucleus is known as Cis face whereas side away from the nucleus is known as Trans face. It is primarily involved in secretion process.

It packages the materials in vesicles and helps in transportation to different cell organelles as well as outside the cell organelle.

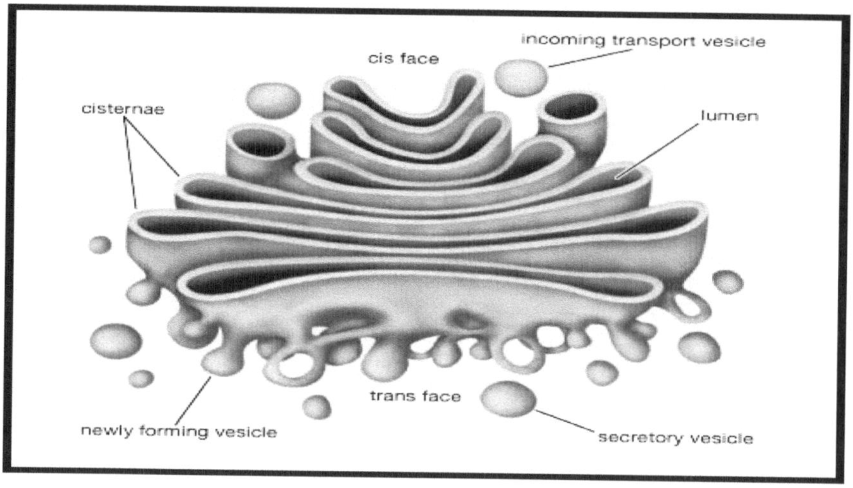

Fig. 6.5. Structure of Golgi apparatus

Lysosomes

It is a membranous structure formed from golgi apparatus. It is garbage collector of the cell. It contains various hydrolytic enzymes such as lipases, proteases which works at acidic environment. They help in digestion of lipids, Proteins and Carbohydrates.

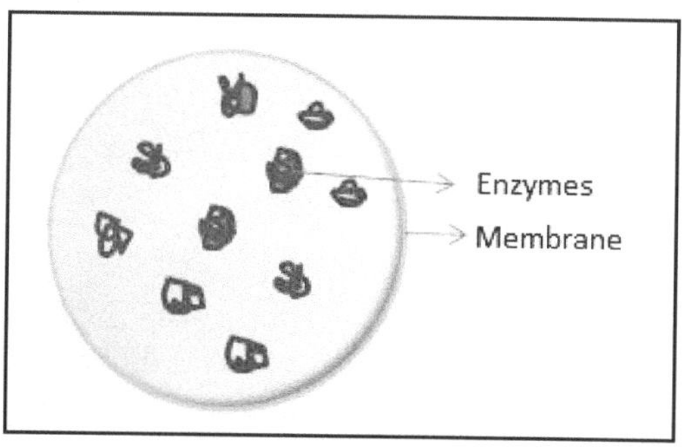

Fig. 6.6. Structure of Lysosomes

Vacuoles

These are more prominent in plant cells as compared to animal cells. It contains water, sap and excretory product. The vacuole is bound by a membrane known as tonoplast. Amoeba contains vacuole - like structure known as contractile vacuole. This is essential for excretion.

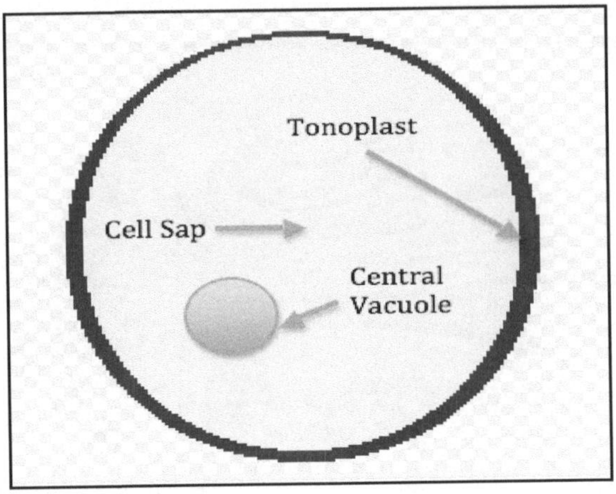

Fig. 6.7. Structure of Vacuole

Mitochondria

It has a double membrane structure *i.e.* the outer membrane and inner membrane. The inner membrane is folded to form a structure known as cristae. It is a site for ATP synthesis. The outer membrane and inner membrane is separated by inter-membrane space. The inner compartment is known as matrix. DNA and ribosomes are present in the matrix of the mitochondria.

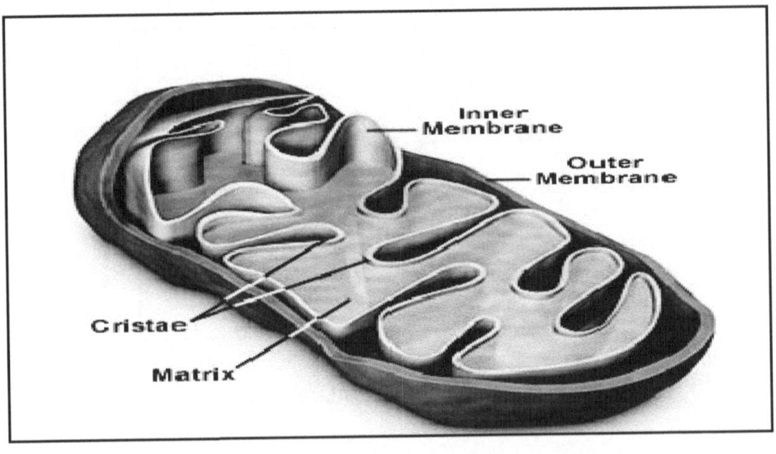

Fig. 6.8. Structure of mitochondria

Plastids

These are found in plant cells. There are three types of plastids *i.e.* chloroplast, chromoplast and leucoplast.

Chloroplast contains a green pigment known as chlorophyll. The leaves appear green due to the presence of chlorophyll in it. The central atom present in chlorophyll is magnesium ion. Chlorophyll is essential for photosynthesis.

Chloroplast is double membrane structure-inner membrane and outer membrane. The space limited by the inner membrane of the chloroplast is called the stroma. Stroma contains flattened sac-like structure known as thylakoids.

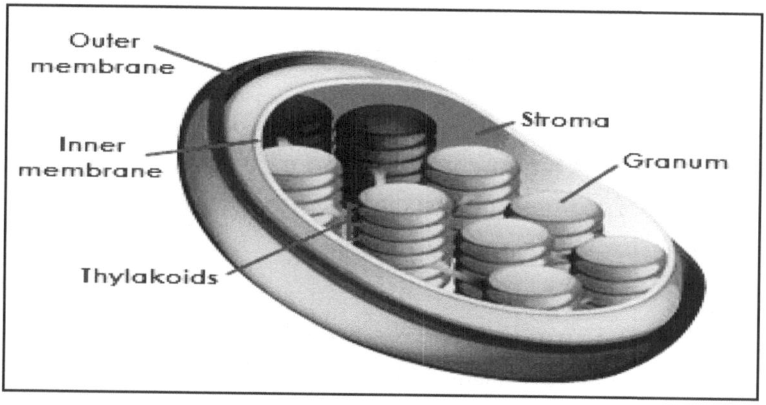

Fig. 6.9. Structure of chloroplast

It is site of photosynthesis. Thylakoids are stacked upon each other to form grana. Each granum is connected with another granum by a lamella. Similar to mitochondria, chloroplast contains DNA and ribosomes.

Chromoplast provides colour to flower and fruits. The pigments present in chromoplast are carotene which provides red colour and xanthophyll which provides yellow-orange colour.

Leucoplast is of different types such as amyloplast for storage of starch, aleuroplast store proteins and elaioplast stores oil and fats.

Nucleus

It is a double membrane structure. The outer membrane is continuous with the ER and contains ribosomes on it. The space between the two membranes is known as perinuclear space.

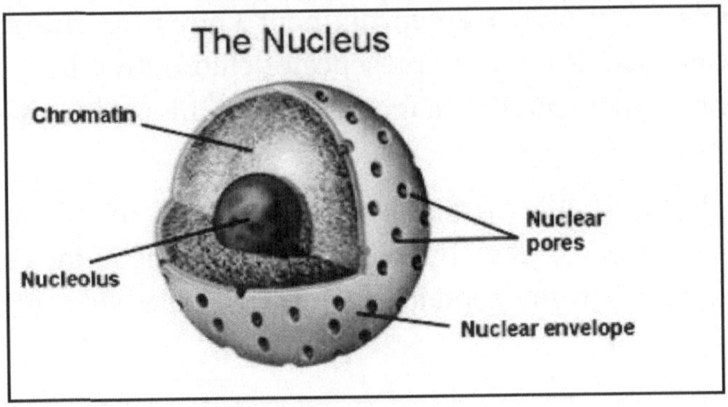

Fig. 6.10. Structure of Nucleus

The two nuclear membranes are interrupted by minute pores known as nuclear pores, which allow only certain molecules of particular size to enter the nucleus. The nuclear matrix is known as nucleoplasm. Nucleoplasm is composed of chromatin and nucleolus.

Chromatin is a network of DNA surrounded by proteins known as histone proteins. During interphase (no cell division), the genetic material exists in the form of chromatin. At the time of cell division, the chromatin becomes compact and forms chromosomes. Chromosomes contain constriction known as centromere. RNA synthesis occurs in the nucleolus.

GLOSSARY

Cell: It is the basic structural, functional and biological unit of all known living organisms.

QUESTIONS

Q. 1. Define cell. Enlist different cell organelles.

Q. 2. Enlist and explain in details different cell organelles along with its function.

Chapter 7

Cell Cycle and Cell Division

INTRODUCTION

'Continuity of life' is an important intrinsic characteristic of living organisms and is achieved through the process of reproduction. The reproduction may be asexual or sexual. Both of these involve the division and replication of cells. Even the growth and development of every living organism depends on the growth and multiplication of its cells. In unicellular organisms, cell division is the means of reproduction by which the mother cell produces two or more new cells. In multi-cellular organism also, new individual develop from a single cell. Cell division is central to life of all cells and is essential for the perpetuation of the species.

It is the process by which a mature cell divides and forms new daughter cells which resemble the parental cell in a number of characters.

Rudolf Virchow (1855 - 1859) was the first to suggest that new cells are formed from the division of the pre-existing cells-omnis cellula e cellula (every cell is derived from a cell). In 1873, Strasburger similarly proposed that nuclei are formed from pre-existing ones. Boveri (1879) and Flemming (1879, 1880) studied details of somatic cell division.

A cell divides when it has grown to a certain maximum size which disturb the karyoplasmic index (KI)/Nucleoplasmic ratio (NP)/Kernplasm connection. Two processes take place during cell reproduction.

- **Cell growth:** (Period of synthesis and duplication of various components of cell).
- **Cell division:** (Mature cell divides into two cells).
- **Cell cycle:** Howard and Pelc (1953) first time described it.

The sequence of events which occur during cell growth and cell division are collectively called cell cycle. Cell cycle completes in two steps:

1. Interphase
2. M-phase/Dividing phase

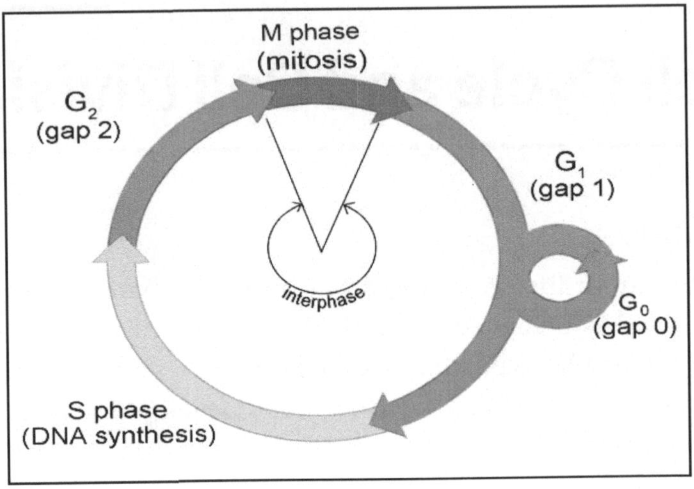

Fig. 7.1. Cell Cycle

1. G_1 - PHASE

It is also called as Post Mitotic/Pre-DNA Synthetic Phase/Gap I[st]. In this phase following events take place:

- Intensive cellular synthesis.
- Synthesis of rRNA, mRNA, ribosomes and proteins.
- Metabolic rate is high.
- Cells become differentiated.
- Synthesis of enzymes and ATP storage.
- Cell size increases.
- Decision for a division in a cell occurs.
- Substances of G stimulate the onset of next S - phase.
- Synthesis of NHC protein, carbohydrates, proteins, lipids.
- Longest and most variable phase.
- Synthesis of enzyme, amino acids, nucleotides etc. but there is no change in DNA amount.

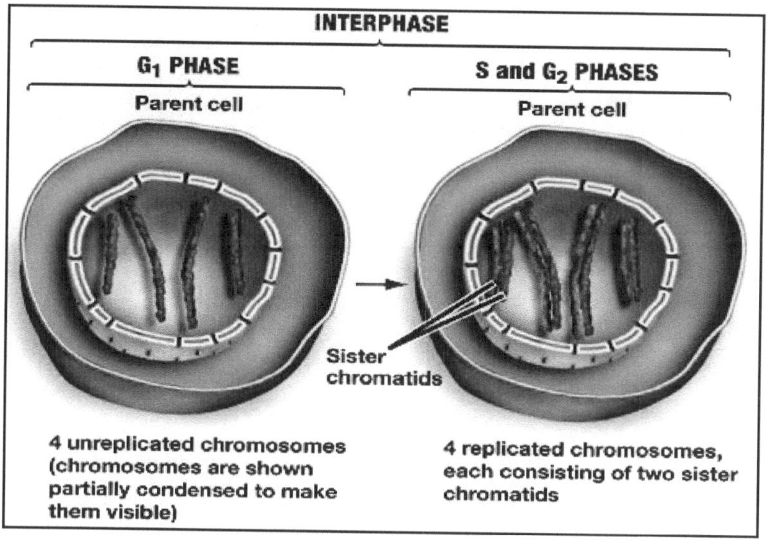

Fig. 7.2. Interphase

2. S - PHASE/SYNTHETIC PHASE

- DNA replicates and its amount becomes double (2C - 4C).
- Synthesis of histone proteins.
- Euchromatin replicates earlier than heterochromatin.
- Synthesis of NHC (non-histone chromosomal proteins).
- Each chromosome has 2 chromatids.

3. G_2 - PHASE

It is also called as Pre Mitotic/Post Synthetic Phase/Gap-IInd. In this phase following events take place:

- Intensive cellular synthesis.
- Increase in energy store.
- Mitotic spindle protein (tubulin) synthesis begins.
- Chromosome condensation factor appears.
- Synthesis of 3 types of RNA and NHC proteins.
- Synthesis of ATP molecule and storage.
- Duplication of mitochondria, plastids and other cellular macromolecular complements.
- Damaged DNA repair occur.

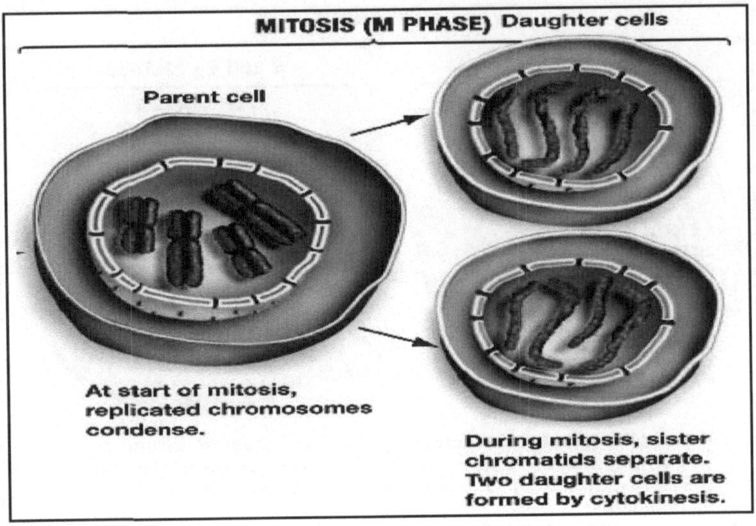

Fig. 7.3. M Phase

4. M - PHASE/DIVIDING PHASE/MITOTIC PHASE

- M - Phase is the final phase of cell cycle.
- It represents the phase of actual division.
- Prior to it, the cell components have undergone duplication.
- M - Phase is the stage of separation of already duplicated components.
- It consists of karyokinesis (division of nucleus) and cytokinesis (division of cell protoplast).

4. G_0 - PHASE (QUIESCENT STAGE)

- It is the stage of inactivation of cell cycle due to non-availability of mitogens and energy rich compounds.
- The cells remain metabolically active.
- They do not grow or differentiate.
- The cells function as reserve cells which can join cell any time.

DIFFERENTIATION PHASE

- Most of the cells leave the G_1 - phase midway.
- They grow in size, assume particular shape and come to have a particular function. The phenomenon is called cell differentiation.

Table 7.1. Differences between G1 and G2 phases

No.	G_1 phase	G_2 phase
1.	It is the first substage of interphase.	It is the last substage of interphase.
2.	Available factors determine its fate, entry in G_0, differentiation or continuity of cell cycle.	There is very little choice for the cell except to proceed further in cell cycle.
3.	Cell organelles do not increase in number.	Cell organelles increase in number.
4.	Cell grows in size but growth of nucleus is little.	Both cell and nucleus grow in size.
5.	It synthesizes RNAs, proteins and other biochemicals for cell growth and subsequent replication of DNA.	It synthesizes RNAs, proteins and other biochemicals for spindle formation and M-phase division.

DURATION OF CELL CYCLE

It depends on the type of cell and external factors such as temperature, food and oxygen. Time period for G_1, S, G_2 and M - phase is species specific under specific environmental conditions. E.g. 20 minutes for bacterial cell, 8-10 hours for epithelial cell, and onion root tip cells may take 20 hours.

REGULATION OF CELL CYCLE

Stage of regulation of cell cycle is G_1 phase during which a cell may follow one of the three options.

- It may start a new cycle, enter the S - phase and finally divide.
- It may be arrested at a specific point of G_1 phase.
- It may stop division and enter G_0 quiescent stage. But when conditions change, cell in G_0 phase can resume the growth and renter the G_1 phase.

Did You Know

- Generation time: Period between 2 successive generations (range 8 hr - 100 days).
- **Mitogens:** Chemicals which enhance or stimulate cell division e.g. lymphokinase (in man)
- Cell cycle duration: 20 minutes in bacteria, 20 hrs in root tip of onion, 2-3 hrs in yeast, 24 hrs in man.
- G_0 phase: Cell only starts dividing when the period is favourable otherwise, it remain viable for months or years as such in G_0 phase.

- During the mitosis of He-La cells, the longest period is gap I phase or G_1.
- DNA replication occurs in S-phase.
- In a cell cycle the condensation of chromosome with visible centromere occurs during M-phase.
- Sequence in cell cycle is G_1, S, G_2 and M.
- M-phase is of shortest duration of cell cycle.
- In G_2, the damaged DNA is repaired.
- Histone protein and RNA synthesis occurs in S - phase.
- Duplication of chromosome occurs at S - phase.

GLOSSARY

Cell cycle: The period in which one cycle of cell division is completed called as cell cycle.

Cell division: It is the process by which a mature cell divides and forms new daughter cells which resemble the parental cell in a number of characters.

Karyokinesis: Division of nucleus.

Cytokinesis: Division of cell protoplast.

QUESTIONS

Q. 1. Define cell cycle and cell division. Enlist and explain phases of cell cycle.

Q. 2. Give the difference between G_1 and G_2 phases of cell cycle.

Test Your Knowledge

Q.1 Which one connected with the cell division

(a) ER (b) Peroxysomes

(c) Ribosomes (d) Microtubules

Q.2 Duplication of chromosomes take place in

(a) G1 phase (b) G2 phase

(c) S phase (d) In all of the above

Q.2 Cell division is initiated by

(a) Centrosome (b) Centriole

(c) Centromere (d) Chromomere

Q.7 The number of DNA in chromosome at G2 stage of cell cycle

(a) One (b) Two

(c) Four (d) Eight

Q.3 Cell division in blue-green algae is more or less similar to that in

(a) Red algae (b) Green algae

(c) Brown algae (d) Bacteria

Q.8 The decision for division occurs in a cell at

(a) S phase (b) G2 phase

(c) G1 phase (d) None of these

Q.4 Which one of the following forms the spindle apparatus during cell division

(a) Chromosome (b) Centrosome

(c) Ribosome (d) Kinetosome

Q.9 During interphase, RNA and proteins are synthesized in

(a) S phase (b) G1 phase

(c) G2 phase (d) In both G1 and G2 phases

Q.5 The replication of nuclear DNA occurs in

(a) G1 phase (b) G2 phase

(c) S phase (d) M phase

Q.10 Mitosis results in

(a) Reduction in chromosome number

(b) Doubling of chromosome number

(c) Constant chromosome number

(d) Increase in cell volume

Chapter 8

Mitosis

INTRODUCTION

The term mitosis derived from Greek word Mitos means thread and osis means state. It is also called indirect cell division or somadtic cell division or equational division. In this, mature somatic cell divides in such a way that chromosomes number is kept constant in daughter cells equal to those in parent cell, so the daughter cells are quantitatively as well as qualitatively similar to the parental cell. So it is called equational division.

Mitosis was first observed by Strasburger (1875) and in animal cell by W. Fleming (1879) term mitosis was given by Fleming (1882). Mitosis is the common method of cell division. It takes place in the somatic cells in the animals. Hence, it is also known as the somatic division. It occurs in the gonads also for the multiplication of undifferentiated germ cells. In plants mitosis occurs in the meristematic cells e.g. root apex and shoot apex. It ranges from 30 minutes to 3 hours time is species-specific but also depends upon type of tissues, temperature.

PROCESS OF MITOSIS

Mitosis is completed in two steps. These are given below

Karyokinesis: (Gk. karyon - nucleus; kinesis - movement) Division of nucleus, Term given by Schneider (1887).

Cytokinesis: (Gk. kitos - cell; kinesis - movement) Division of cytoplasm, Term given by Whitemann (1887).

1. KARYOKINESIS

It comprises four phases *i.e.* Prophase, Metaphase, Anaphase & Telophase.

PROPHASE

It is largest phase of karyokinesis.

- Chromatin fibres thicken and shorter to form chromosomes which may overlap each other and appears like a ball of wool. *i.e.* Spireme stage.
- Each chromosome divides longitudinally into 2 chromatids which remain attached to centromere.
- Nuclear membrane starts disintegrating except in dinoflagellates.
- Nucleolus starts disintegrating.
- Cells become viscous, refractive and oval in outline.
- Spindle formation begins.
- Cell cytoskeleton, Golgi complex, ER etc. disappear.
- In animal cells, centrioles move towards opposite sides.
- Lampbrush chromosomes can be studied well.
- Small globular structures (beaded) on the chromosome are called chromomeres.

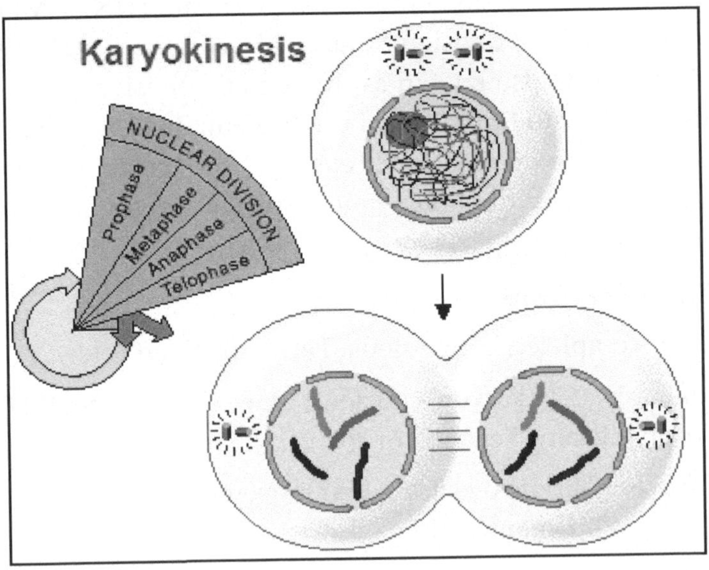

Fig. 8.1. Karyokinesis

METAPHASE

- Chromosomes become maximally distinct i.e. size can be measured.
- A colourless, fibrous, bipolar spindle appears.
- Spindle is formed from centriole (in animal cells) or MTOC (microtubule organising centre) in plant cells successively called astral and anastral spindle.
- Spindle has 3 types of fibres.
  - Continuous fibre (run from pole to pole).
  - Discontinuous fibre (run between poles to centromeres).
  - Interzonal fibre (run between 2 centromere).
- Spindle fibres are made up of 97% tubulin protein and 3% RNA.
- Chromosomes move towards equatorial plane of spindles called congression and become arranged with their arms directed towards pole and centromere towards equator.
- Spindle fibres attach to kinetochores.
- Metaphase is the best stage for studying chromosome morphology.

ANAPHASE

- Centromere splits from the middle and two chromatids get separated.
- Both the chromatids move towards opposite poles due to repulsive force called anaphasic movement.
- Anaphasic movement is brought about by the repolymerisation of continuous fibres and depolymerisation of chromosomal fibres.
- Different shape of chromosomes becomes evident during chromosome movement viz. metacentric acrocentric etc.
- A chromosome takes V, J, I or L shapes.
- The centromere faces towards equator.
- The chromatids are moved towards the pole at a speed of 1 mm/ minute. About 30 ATP molecules are used to move one chromosome from equator to pole.

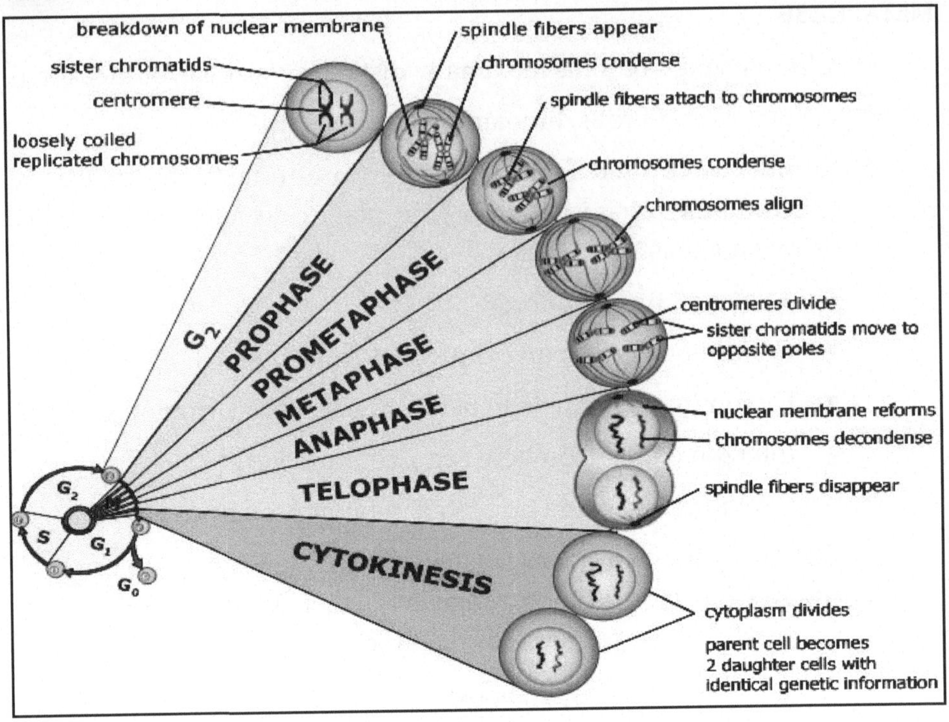

Fig. 8.2. Phases of Mitosis

TELOPHASE

- ❖ Chromosomes reached on poles by the spindle fibers and form two groups.
- ❖ Chromosomes begin to uncoil and form chromatin net.
- ❖ The nuclear membrane and nucleolus reappear.
- ❖ Two daughter nuclei are formed.
- ❖ Golgi complex and ER etc., reform.

2. CYTOKINESIS

It involves division of cytoplasm in animal cells, the cell membrane develops a constitution which deepens centripetally and is called cell furrow method. In plant cells, cytokinesis occurs by cell plate formation.

SIGNIFICANCE OF MITOSIS

- ❖ It keeps the chromosome number constant and genetic stability in daughter cells, so the linear heredity of an organism is maintained. All the cells are with similar genetic constituents.

- It helps in growth and development of zygote into adult through embryo formation.
- It provides new cells for repair and regeneration of lost parts and healing of the wounds.

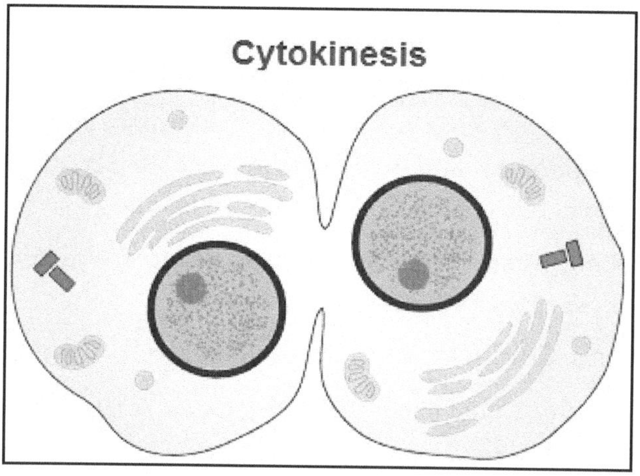

Fig. 8.3. Cytokinesis

- It helps in asexual reproduction by fragmentation, budding, stem cutting, etc.
- It also restores the nucleo - plasmic ratio.
- Somatic variations when maintained by vegetative propagation can play important role in speciation.

Did You Know

- Pericentriolar cloud: A clear cytoplasmic area with no cell organelle between the centriole pair and astral rays.
- Root tips of onion are best material for studying mitosis.
- Kinetochore: A discoidal area on each chromatid and is the site of attachment of spindle fibres.
- In mitosis, plectonemic coiling takes place, in which sister chromatids are tightly coiled upon each other and are not easily separable. Paranemic coiling found in meiosis.
- Chromosomal fibres are also called tractile fibres, while continuous fibres are also called interpolar fibres.

- Mitogens: The agents which stimulate cell division e.g., cytokinins, auxins, gibberllins, insulin, temperature, steroids.
- Mitotic poision: The agents which inhibit cell division.
- Azides and Cyanides: Inhibit prophase.
- Colchicine: Inhibits spindle formation at metaphase.
- Mustard gas: Agglutinates the chromosomes.
- Chalones: These were first reported by Laurence and Bullough (1960). They are peptides and glycoproteins secreated by extracellular fluid of healthy cells and inhibit cellular division.
- Karyochoriosis: A type of mitosis in fungi in which is intranuclear nucleus divides by furrow formation.
- C-mitosis: Colchicine induced mitosis.
- After undergoing certain divisions, cells die. This is called as "Hayflick limit".
- Actinomycin D and tetracyclin inhibit cell division.
- 7 mitotic divisions occur to form embryosac in angiosperms.
- Mitosis index is the ratio of dividing and non-dividing cells.

QUESTIONS

Q. 1. What do you mean by mitosis? Enlist and explain stages of mitosis along with neat labelled diagram.

Q. 2. Define mitosis. What is the significance of mitosis?

Chapter 9

Meiosis

INTRODUCTION

The term meiosis derived from Greek word Meioum means to reduce and osis means state. It is a special type of division in which the chromosomes duplicate only once, but cell divides twice. So one parental cell produces four daughter cells; each having half the chromosome number and DNA amount than normal parental cell. So meiosis is also called reductional division.

It was first demonstrated by Van Benden (1883) but was described by Winiwarter (1900). Term 'meiosis' was given by Farmer and Moore (1905). It is found in special types and at specific period. It is reported in diploid germ cells of sex organs (e.g. primary spermatocytes of testes to form male gametes called spermotozoa and primary oocytes to form female gametes called ova in animals) and in pollen mother cells (microsporocytes) of anther and megasporocyte of ovule of ovary of flowers in plant to form the haploid spores. The study of meiosis in plants can be done in young flower buds.

PROCESS OF MEIOSIS

Meiosis is completed in two steps, meiosis I and meiosis II

MEIOSIS I

In which the actual chromosome number is reduced to half. Therefore, meiosis I is also known as reductional division or heterotypic division. It results in the formation of two haploid cells from one diploid cell. It is divided into two parts, karyokinesis I and cytokinesis I.

KARYOKINESIS I

It involves division of nucleus. It is divided into four phases i.e. prophase, metaphase, anaphase and telophase.

PROPHASE I

It is of longest phase of karyokinesis of meiosis. It is again divisible into five subphases i.e. leptotene, zygotene, pachytene, diplotene and diakinesis.

1. LEPTOTENE/LEPTONEMA

- Chromosomes are long thread like with chromomeres on it.
- Volume of nucleus increases.
- Chromatin network has half chromosomes from male and half from female parent.
- Chromosomes with similar structure are known as homologous chromosomes.
- Leptonemal chromosomes have a definite polarization and forms loops whose ends are attached to the nuclear envelope at points near the centrioles, contained within an aster. Such peculiar arrangement is termed as bouquet stage (in animals) and syndet knot (in plants).
- Electron microscope reveals that chromosomes are composed of paired chromatids, a dense proteinaceous filament or axial core lies within the groove between the sister chromatids of each chromosome.
- Lampbrush chromosome found in oocyte of amphibians is seen in leptotene.

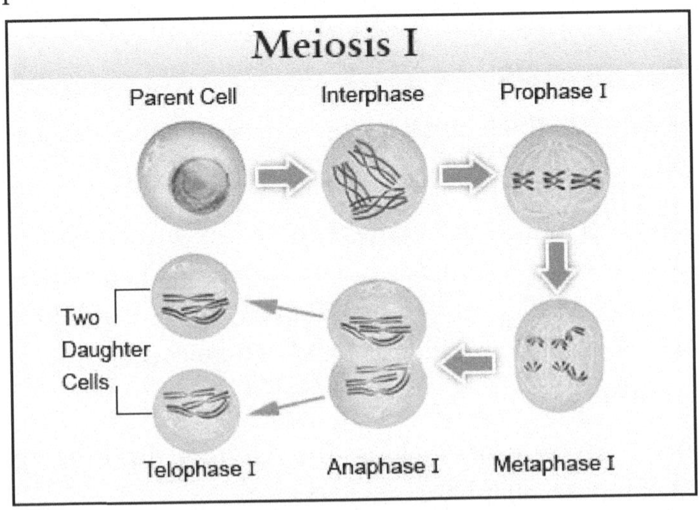

Fig. 9.1. Meiosis I

2. ZYGOTENE/ZYGONEMA

- Pairing or 'synapsis' of homologous chromosomes takes place in this stage.
- Synapsis may be of following types.
  - Procentric: Starting at the centromere.
  - Proterminal: Starting at the end.
  - Localised random: Starting at various points.
- Paired chromosomes are called bivalents, which by further molecular packing and spiralization becomes shorter and thicker.
- Pairing of homologous chromosomes in a zipper-fashion. Number of bivalents (paired homologous chromosomes) is half to total number of chromosomes in a diploid cell. Each bivalent is formed of one paternal and one maternal chromosome (i.e. one chromosome derived from each parent).
- Under Electron Microscope, a filamentous ladder like nucleoproteinous complex, called synaptinemal. Complex between the homologous chromosomes which is discovered by "Moses" (1953).

4. PACHYTENE/PACHYNEMA

- In the tetrad, two similar chromatids of the same chromosome are called sister chromatids and those of two homologous chromosomes are termed non-sister chromatids.
- Crossing over i.e. exchange of segments between non-sister chromatids of homologous chromosome occurs at this stage.
- It takes place by breakage and reunion of chromatis segments. Breakage called nicking is assisted by an enzyme endonuclease and reunion termed annealing is added by an enzyme ligase. Breakage and reunion hypothesis proposed by Darlington (1937).
- Chromatids of pachytene chromosome are attached with centromere.
- A tetrad consists of two sets of homologous chromosomes each with two chromatids. Each tetrad has four kinetochore (two sister and two homologous).

- A number of electron dense bodies about 100 nm in diameter are seen at irregular intervals within the centre of the synaptonemal complex, known as recombination nodules.
- DNA polymerase is responsible for the repair synthesis.

5. DIPLOTENE/DIPLONEMA

- At this stage the paired chromosomes begin to separate (desynapsis).
- Cross is formed at the place of crossing over between non-sister chromatids.
- Homologous chromosomes move apart they remain attached to one another at specific points called chiasmata.
- At least one chiasma is formed in each bivalent.
- Chromosomes are attached only at the place of chiasmata.
- Chromatin bridges are formed in place of synaptonemal complex on chiasmata.
- This stage remains as such for long time.

6. DIAKINESIS

- Chiasmata moves towards the ends of chromosomes. This is called terminalization.
- Chromatids remain attached at the place of chiasma only.
- Nuclear membrane and nucleolus degenerates.
- Chromosome recondenses and tetrad moves to the metaphase plate.
- Formation of spindle.
- Bivalents are irregularly and freely scattered in the nucleocytoplasmic matrix.

When the diakinesis of prophase-I is completed than cell enters into the metaphase I.

METAPHASE I

- Chromosome comes on the equator.
- Due to repulsive force the chromosome segment get exchanged at the chiasmata.
- Bivalents arrange themselves in two parallel equatorial or metaphase plates. Each equatorial plate has one genome.

- Centromeres of homologous chromosomes lie equisdistant from equator and are directed towards the poles while arms generally lie horizontally on the equator.
- Each homologous chromosome has two kinetochores and both the kinetochores of a chromosome are joined to the chromosomal or tractile fibre of same side.

ANAPHASE I

- It involves separation of homologous chromosomes which start moving opposite poles so each tetrad is divided into two daughter dyads. So anaphase I involves the reduction of chromosome number, this is called disjunction.
- The shape of separating chromosomes may be rod or J or V shape depending upon the position of centromere.
- Segregation of Mendelian factors or independent assortment of chromosomes takes place. In which the paternal and maternal chromosomes of each homologous pair segregate during anaphase-I which introduces genetic variability.

TELOPHASE I

- Two daughter nuclei are formed but the chromosome number is half than the chromosome number of mother cell.
- Nuclear membrane reappears.
- After telophase I, cytokinesis may or may not occur.
- At the end of Meiosis I either two daughter cells will be formed or a cell may have two daughter nuclei.
- Meiosis I is also termed as reduction division.
- After meiosis I, the cells in animals are reformed as secondary spermatocytes or secondary oocytes; with haploid number of chromosomes but diploid amount of DNA.
- Chromosomes undergo decondensation by hydration and despiralization and change into long and thread like chromatin fibres.

INTERPHASE

Generally there is no interphase between meiosis-I and meiosis-II. A brief interphase called interkinesis or intermeiotic interphase. There is no replication of chromosome, during this interphase.

CYTOKINESIS I

It may or may not be present. When present, it occurs by cell-furrow formation in animal cells and cell plate formation in plant cells.

SIGNIFICANCE OF MEIOSIS I

- It separates the homologous chromosomes to reduce the chromosome number to the haploid state, a necessity for sexual reproduction.
- It introduces variation by forming new gene combinations through crossing over and randon assortment of paternal and maternal chromosomes.
- It may at times cause chromosomal mutation by abnormal disjunction.
- It induces the cells to produce gametes for sexual reproduction or spores for asexual reproduction.

MEIOSIS II

It is also called equational or homotypical division because the number of chromosomes remains same as after meiosis-I. It is of shorter duration than even typical mitotic division. It is also divisible into two parts, Karyokinesis-II and Cytokinesis-II.

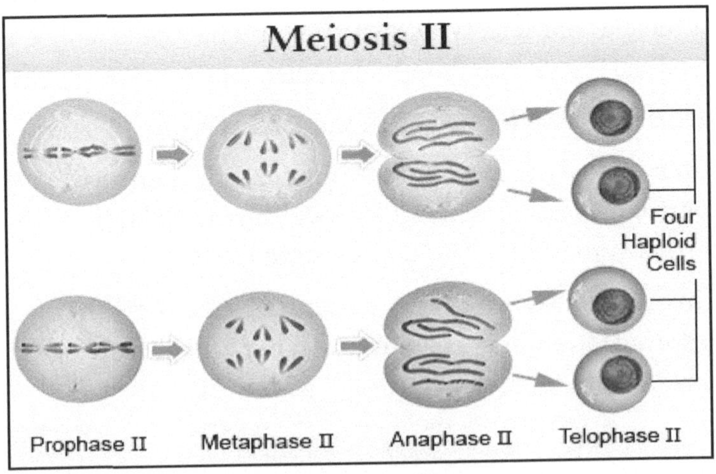

Fig. 9.2. Meiosis II

KARYOKINESIS II

It involves the separation of two chromatids of each chromosome and their movement to separate cells. It is divided in four phases *i.e.,* Prophase II, Metaphase II, Anaphase II and Telophase II. Almost all the changes of Karyokinesis II resembles to mitosis which involves.

- It starts just after end of telophase I.
- Each daughter cell (nucleus) undergoes mitotic division.
- It is exactly similar to mitosis.
- At the end of process, cytokinesis takes place.
- Four daughter cells are formed after completion.
- The sister kinetochores of one chromosome are separated.
- The four daughter cells receive one chromatid each of the tetravalent.
- Centromere divides at anaphase II.
- Spindle fibres contract at prophase II.

CYTOKINESIS II

It is always present and occurs by cell furrow formation in animal cell and cell plate formation in plant cell.

So by meiosis, a diploid parental cell divides twice forming four haploid gametes or sex cells, each having half the DNA amount than that of the parental cell and one-fourth of DNA present in the cell at the time of beginning of meiosis.

SIGNIFICANCE OF MEIOSIS

- Constancy of chromosome number in successive generation is brought by process.
- Chromosome number becomes half during meiosis.
- It helps in introducing variations and mutation.
- It brings about gamete formation.
- It maintains the amount of genetic informative material.
- Sexual reproduction includes one meiosis and fusion.
- The four daughter cells will have different types of chromatids.

NECESSITY OF MEIOSIS II

The basic aim of meiosis is to reduce the number of chromosomes to half. The chromosomes that separate in the anaphase of meiosis I are still double. Each consists of two chromatids and has 2n amount of DNA. Thus reduction of DNA content does not occur in meiosis I. Truly haploid nuclei in terms of DNA contents as well as chromosome number are formed in meiosis II. When the chromatids of each chromosome are separated into different nuclei, thus meiosis II is necessary.

Table 9.1. Difference between Mitosis and Meiosis

No.	Characters	Mitosis	Meiosis
I. General			
(1)	Site of occurrence	Somatic cells and during the multiplicative phase of gametogenesis in germ cells.	Reproductive germ cells of gonads.
(2)	Period of occurrence	Throughout life.	During sexual reproduction.
(3)	Nature of cells	Haploid or diploid.	Always diploid.
(4)	Number of divisions	Parental cell divides once.	Parent cell divides twice.
(5)	Number of daughter cells	Two.	Four.
(6)	Nature of daughter cells	Genetically similar to parental cell. Amount of DNA and chromosome number is same as in parental cell.	Genetically different from parental cell. Amount of DNA and chromosome number is half to that of parent cell.
II. Prophase			
(7)	Duration	Shorter (of a few hours) and simple.	Prophase I is very long (may be in days or months or years) and complex.
(8)	Sub phases	Formed of 3 sub phases: early-prophase, mid-prophase and late-prophase.	Prophase I is formed of 5 subphases: leptotene, zygotene, pachytene, diplotene and diakinesis.
(9)	Bouquet stage	Absent.	Present in leptotene stage.
(10)	Synapsis	Absent.	Pairing of homologous chromosomes in zygotene stage.
(11)	Chiasma formation and crossing over.	Absent.	Occurs during pachytene stage of prophase I.
(12)	Disappearance of nucleolus and nuclear membrane	Comparatively in earlier part.	Comparatively in later part of prophase-I.
(13)	Nature of coiling	Plectonemic.	Paranemic.
III. Metaphase			
(14)	Metaphase plates	Only one equatorial plate	Two plates in metaphase I but one plate in metaphase II.

No.	Characters	Mitosis	Meiosis
(15)	Position of centromeres	Lie at the equator. Arms are generally directed towards the poles.	Lie equidistant from equator and towards poles in metaphase I while lie at the equator in metaphase II.
(16)	Number of chromosomal fibres	Two chromosomal fibres join at centromere.	Single in metaphase I while two in metaphase II.
IV. Anaphase			
(17)	Nature of separating chromosomes	Daughter chromosomes (chromatids with independent centromeres) separate.	Homologous chromosomes separete in anaphase I while chromatids separate in anaphase in anaphase II.
(18)	Splitting of centromeres and development of inter-zonal fibres	Occurs in anaphase.	No splitting of centromeres. Inter-zonal fibres are developed in metaphase I.
V. Telophase			
(19)	Occurrence	Always occurs	Telophase I may be absent but telophase II is always present.
VI. Cytokinesis			
(20)	Occurrence	Always occurs	Cytokinesis I may be absent but cytokinesis II is always present.
(21)	Nature of daughter cells	2n amount of DNA than 4n amount of DNA in parental cell.	1n amount of DNA than 4n amount of DNA in parental cell.
(22)	Fate of daughter cells	Divide again after interphase.	Do not divide and act as gametes.
VII. Significance			
(23)	Functions	Helps in growth, healing, repair and multiplication of somatic cells. Occurs in both asexually and sexually reproducing organisms.	Produces gametes which help in sexual reproduction.
(24)	Variations	Variations are not produced as it keeps quality and quantity of genes same.	Produces variations due to crossing over and chance arrangement of bivalents at metaphase I.
(25)	In evolution	No role in evolution.	It plays an important role in speciation and evolution.

Did You Know

- Brachymeiosis: Failure of meiosis-II. It is characteristic feature of fungi.
- Meiosis II is not mitosis as it occurs haploid number of chromosomes and chromatids formed may not be similar to each other.

- ❖ Restitution nucleus: A colchicine treated cell has the nucleus with double sets of chromosomes.
- ❖ In cyperus, one meiosis produce only one pollen instead of four so that meiotic division required producing fruits will be - number of fruits 2.
- ❖ Chiasmata first observed by Janseens (1909).
- ❖ When sister chromatids are loosely arranged and are easily separate. It is found in meiotic chromosomes.
- ❖ Mitosis ends in 1-2 hours while meiosis may take 24 hrs to few years.
- ❖ Neuron cells are always in interphase.
- ❖ When the chromosome duplicates but karyokinesis does not take place the number of chromosome per cell will increases, it is called endomitosis or endoduplication.
- ❖ Process of inducing mitosis into a cell - mitogenesis.
- ❖ To study mitosis root tips are fixed in 1: 3 acetic acid and methanol.
- ❖ *Colchicine* inhibits spindle formation and enhances duplication in number of chromosomes.
- ❖ At the time of cell division electrostatic force is responsible for terminalization.
- ❖ Mitotic crossing over takes place in parasexual cycle.

QUESTIONS

Q. 1. What do you mean by meiosis? Enlist phases of meiosis.

Q. 2. Enlist and explain the phases of meiosis with neat labelled diagrams.

Q. 3. What are the significances of the meiosis?

Q. 4. Give the difference between meiosis and mitosis.

Test Your Knowledge

Q. 1 Meiosis was discovered by

(a) Strasburger (b) Hofmeister

(c) Sutton (d) Amici

Q. 2 Repulsion of homologous chromosomes takes place in

(a) Zygotene (b) Leptotene

(c) Diakinesis (d) Diplotene

Q. 3 In which of the following meiosis takes place

(a) Pollen grains (b) Pollen tube

(c) Pollen mother cells (d) Generative cells

Q. 4 Chiasmata formation occurs during

(a) Diplotene (b) Leptotene

(c) Pachytene (d) Diakinesis

Q. 5 The significance of meiosis lies in

(a) Reduction of the diploid number of chromosomes to haploid

(b) Maintaining constancy in the number of diploid chromosomes during sexual reproduction

(c) Production of genetic variability in the population of a species

(d) All of the above

Q. 6 In pachytene stage of meiosis the chromosomes appear

(a) Single stranded (b) Double stranded

(c) Three stranded (d) Four stranded

Q.9 Pachytene occurs during

(a) Meiosis (b) Mitosis

(c) Growth of a cell (d) Formation of endosperm

Q. 7 When during the meiotic division, do the homologous chromosomes pair with each other

(a) Leptotene (b) Pachytene

(c) Zygotene (d) Metaphase I

Q. 8 Chromonemata start associating into bivalent chromosomes during

(a) Zygotene (b) Leptotene

(c) Pachytene (d) Diplotene

Q. 9 The correct chronological sequence Prophase of reduction division is

(a) Leptotene - pachytene - zygotene - diplotene - diakinesis

(b) Leptotene - diplotene - pachytene - zygotene - diakinesis

(c) Leptotene - zygotene - diplotene - pachytene - diakinesis

(d) Leptotene - zygotene - pachytene – diplotene - diakinesis

Chapter 10

Introduction to Genetics

INTRODUCTION

Genetics is a biological science which deals with principles of heredity and variation. Heredity refers to the transmission of characters from parents to offspring's. Thus, Genetics is a science which unravels the inheritance of various traits from one generation to next generation. The foundation of this new branch of biology was laid by Mendel in 1866 when he discovered the principles of heredity. However, Mendel's findings came into light only in 1900 when similar results were obtained independently by three scientists, viz., De Vres, Carl Correns and Tschermak. Thus, genetics was born in 1900.

Genetics is the branch of botany which deals with genes, heredity and variation in living organisms. The term genetics was first used by Bateson in 1905 *i.e.* five year after, after its birth. The word genetics has been derived from the Greek word gene means to become or to grow.

BRANCHES OF GENETICS

- **Plant genetics:** It deals with inheritance of traits in various plant species.
- **Animal genetics:** It deals with inheritance of traits in animals.
- **Microbial genetics:** It deals with inheritance of traits in micro-organisms like bacteria, viruses & fungi.
- **Cytogenetics:** It deals with combined study of cytology and genetics.
- **Biochemical Genetics:** It deals with the role of genes in controlling biochemical pathways in an organism.
- **Molecular Genetics:** It deals with the structure, composition, function & replication of chromosomes and genes, representing genetic material. viz., DNA & RNA.

- **Radiation Genetics:** It deals with effect of various types of raditions on chromosomes and genes.
- **Euphenics:** It deals with the control of hereditary diseases especially inborn errors of metabolism.
- **Mendelian Genetics:** It deals with inheritance of qualitative traits or oligogenic traits which display discontinuous variation.
- **Population Genetics:** It deals with inheritance of quantitative traits or polygenic traits which display continuous variation. This branch of genetics also called as biometrical genetics or statistical genetics or mathematical genetics.

EUGENICS

It is branch of botany in which application of principles of genetics for improving the human being or races.

The science of improving a population by controlled breeding to increase the occurrence of desirable heritable characteristics.

EVOLUTIONARY GENETICS

Evolutionary genetics is defined as how genetic variation leads to evolutionary change. It includes, evolution of structure of genome, genetic basis of adaptation and specification and genetic changes in response to selection within the population. Evolutionary genetics is the sum of population genetics and evolution.

GENOME EVOLUTION

Genome evolution is defined as the process by which genome changes in structure and size with time.

PRE-MENDELIAN THEORIES OF HEREDITY

Some of the various theories of heredity about heredity which were proposed by various scientists prior to the discovery of Mendel are briefly given below:

1. Preformation Theory

This theory was proposed by two Dutch biologists, Swammerdam and Bonnet (1720-1793). This theory states that a miniature human, called homunculus, is already present in the egg and sperm which grows into human being after formation of zygote. This theory was soon rejected because this could not be proved scientifically.

2. Theory of Epigenesis

Proposed by a German biologist Wolff (1738-1794), the theory states that the egg and sperms are undifferentiated cells. The differentiation into various organs/parts takes place only after fertilization in the zygote resulting in the development of adult tissues and organs. This concept is universally accepted.

3. Theory of Acquired Characters

Proposed by the famous French biologist Lamarck (1744-1829) this theory states that a new character once acquired by an individual shall pass on to its progeny. It means if a man develops a strong muscle by exercise, all his children would have strong muscles. On the other hand, if a person becomes weak all his children would be weak. This concept was totally rejected by Weismann on the basis of experiment on rats performed through 22 generations.

4. Theory of Pengenes

This theory, proposed by the famous English naturalist Charles Darwin (1809-1882), states that very small, exact but invisible copies of each body organ and component (called gemmules) are transported by the blood stream to the sex organs. These gemmules are assembled in the gametes. After fertilization these gemmules move out to different parts of the body resulting in the development of respective organ. A defective gemmule will lead to the development of defective organ in an individual. This theory too was rejected because it lacked scientific basis.

5. Germplasm Theory

This theory, advocated by August Weismann (1889), a German biologist, states that body tissues are of two types, viz., germplasm and somatoplasm. The germplasm refers to the reproductive tissues or cells which produce gametes.

The somatoplasm includes all other body tissues which are not related to sexual reproduction. Thus, transmission of characters from one generation to other takes place only through germplasm. Any change in the germplasm will lead to change in the next generation. This theory is accepted in a broad sense.

MENDEL'S EXPERIMENT

For thousands of years farmers and herders have been selectively breeding their plants and animals to produce more useful hybrids. It was somewhat of a hit or miss process since the actual mechanisms governing inheritance were unknown. Knowledge of these genetic mechanisms finally came as a result of careful laboratory breeding experiments carried out over the last century and a half.

G. J. Mendel (1822-1884)

By the 1890's, the invention of better microscopes allowed biologists to discover the basic facts of cell division and sexual reproduction. The focus of genetics research then shifted to understanding what really happens in the transmission of hereditary traits from parents to children. A number of hypotheses were suggested to explain heredity, but Gregor J. Mendel, a little known Central European monk, was the only one who got it more or less right. His ideas had been published in 1866 but largely went unrecognized until 1900, which was long after his death. His early adult life was spent in relative obscurity doing basic genetics research and teaching high school mathematics, physics and Greek in Brno (now in the Czech Republic). In his later years, he became the abbot of his monastery and put aside his scientific work.

While Mendel's research was with plants, the basic underlying principles of heredity that he discovered also apply to people and other animals because the mechanisms of heredity are essentially the same for all complex life forms.

Through the selective cross-breeding of common pea plants (*Pisum sativum*) over many generations, Mendel discovered that certain traits show up in offspring without any blending of parent characteristics. For instance, the flowers of pea are either purple or white-intermediate colours do not appear in the offspring of cross-pollinated pea plants. Mendel observed seven traits that are easily recognized and apparently only occur in one of two forms, which are given below

Table 10.1. Characters of pea plant studied by Mendel

Sr. No.	Name of the traits
1.	Flower colour is purple or white
2.	Flower position is axil or terminal
3.	Stem length is long or short
4.	Seed shape is round or wrinkled
5.	Seed colour is yellow or green
6.	Pod shape is inflated or constricted
7.	Pod colour is yellow or green

This observation that these traits do not show up in offspring plants with intermediate forms was critically important because the leading theory in biology at the time was that inherited traits blend from generation to generation. Most of the leading scientists in the 19th century accepted this 'blending theory'. Charles Darwin proposed another equally wrong theory known as 'Pangenesis'. This held that hereditary particles in our bodies are affected by the things we do during our lifetime. These modified particles were thought to migrate via blood to the reproductive cells and subsequently could be inherited by the next generation. This was essentially a variation of Lamarck's incorrect idea of the 'Inheritance of acquired characteristics'.

Mendel picked common garden pea plants for the focus of his research because they can be grown easily in large numbers and their reproduction can be manipulated. Pea plants have both male and female reproductive organs. As a result, they can either self-pollinate themselves or cross-pollinate with another plant. In his experiments, Mendel was able to selectively cross-pollinate purebred plants with particular traits and observe the outcome over many generations. This was the basis for his conclusions about the nature of genetic inheritance.

In cross-pollinating plants that either produces yellow or green pea seeds exclusively, Mendel found that the first offspring generation (F_1) always has yellow seeds. However, the following generation (F_2) consistently has a 3:1 ratio of yellow to green.

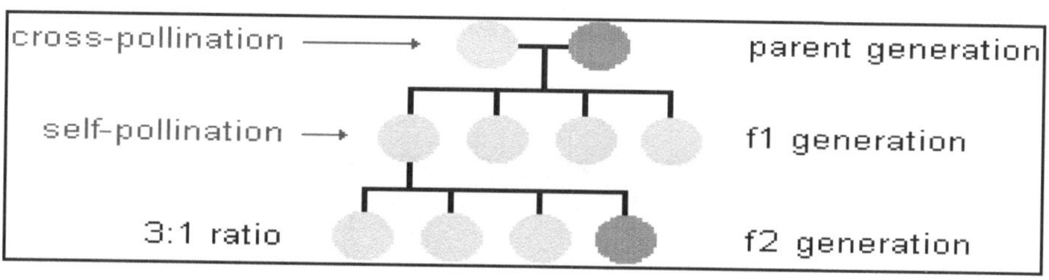

This 3:1 ratio occurs in later generations as well. Mendel realized that this underlying regularity was the key to understanding the basic mechanisms of inheritance.

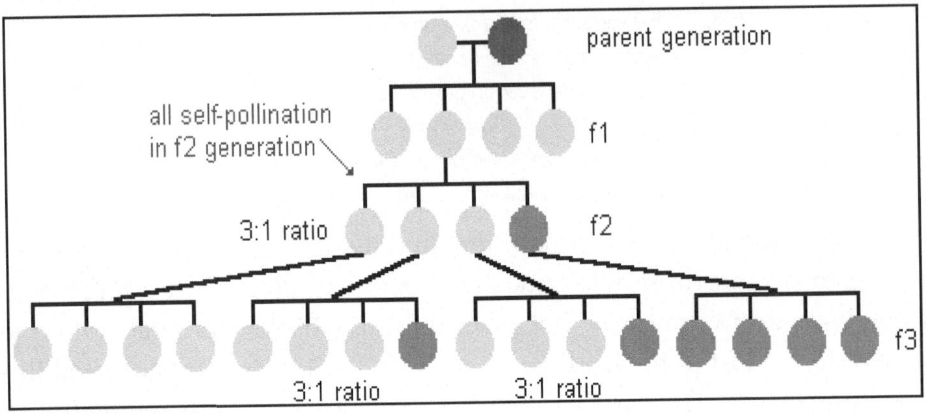

He came to three important conclusions from these experimental results:

1.	That the inheritance of each trait is determined by 'units' or 'factors' that are passed on to descendants unchanged. (these units are now called **genes**)
2.	That an individual inherits one such unit from each parent for each trait.
3.	That a trait may not show up in an individual but can still be passed on to the next generation.

It is important to realize that, in this experiment, the starting parent plants were homozygous for pea seed colour. That is to say, they each had two identical forms (or alleles) of the gene for this trait - 2 yellows or 2 greens. The plants in the F1 generation were all heterozygous. In other words, they each had inherited two different alleles-one from each parent plant. It becomes clearer when we look at the actual genetic makeup, or genotype, of the pea plants instead of only the phenotype, or observable physical characteristics.

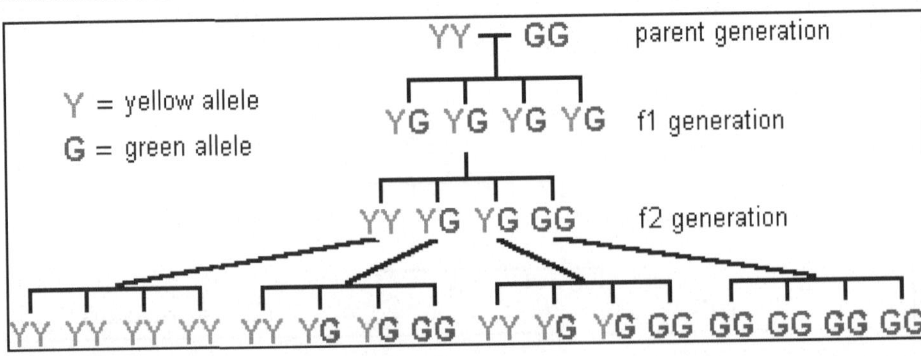

Note that each of the F_1 generation plants (shown above) inherited a Y allele from one parent and a G allele from the other. When the F_1 plants breed, each has an equal chance of passing on either Y or G alleles to each offspring.

With all of the seven pea plant traits that Mendel examined, one form appeared dominant over the other, which is to say it masked the presence of the other allele. For example, when the genotype for pea seed colour is YG (heterozygous), the phenotype is yellow. However, the dominant yellow allele does not alter the recessive green one in any way. Both alleles can be passed on to the next generation unchanged.

Mendel's observations from these experiments can be summarized in two principles

1.	The law of segregation (Mendels' first law or principle)
2.	The law of independent assortment (Mendels' second law or principle)

According to the law of segregation, for any particular trait, the pair of alleles of each parent separate and only one allele passes from each parent on to an offspring. Which allele in a parent's pair of alleles is inherited is a matter of chance. We now know that this segregation of alleles occurs during the process of sex cell formation (i.e., meiosis).

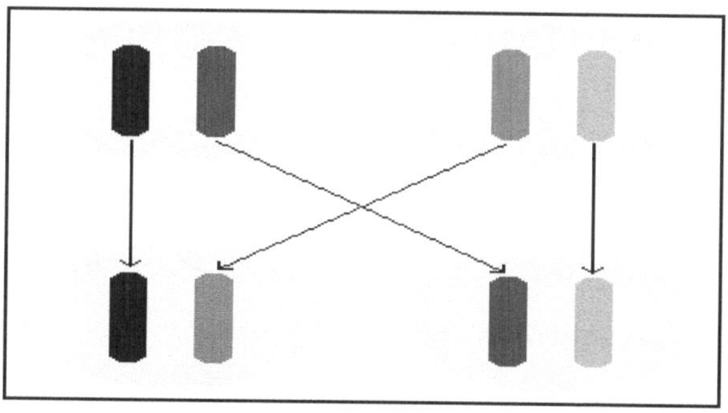

Fig. 10.1. Segregation of alleles in the production of sex cells

According to the law of independent assortment, different pairs of alleles are passed to offspring independently of each other. The result is that new combinations of genes present in neither parent are possible. For example, a pea plant's inheritance of the ability to produce purple flowers instead of white ones does not make it more likely that it will also inherit the ability to produce yellow pea seeds in contrast to green ones. Likewise, the principle of independent assortment explains why the human inheritance of a particular eye colour does not increase or decrease the likelihood of having 6 fingers on each hand. Today, we know this is due to the fact that the genes for independently assorted traits are located on different chromosomes.

These two principles of inheritance, along with the understanding of unit inheritance and dominance, were the beginnings of our modern science of genetics. However, Mendel did not realize that there are exceptions to these rules. Some of these exceptions will be explored in the third section of this tutorial and in the Synthetic Theory of Evolution tutorial.

By focusing on Mendel as the father of genetics, modern biology often forgets that his experimental results also disproved Lamarck's theory of the inheritance of acquired characteristics described in the Early Theories of Evolution tutorial. Mendel rarely gets credit for this because his work remained essentially unknown until long after Lamarck's ideas were widely rejected as being improbable.

GLOSSARY

Genetics: A biological science which deals with the principles of heredity and variation.

Heredity: Transmission of traits from parents to offspring.

Variation: Differences for various traits among the individuals of the same species.

Gene: It is the smallest segment of DNA which governs one trait.

Allele: Alternative forms of gene known as allele.

Genotype: Genetic constitution or make up of an organism.

Phenotype: External appearance of an organism.

Dominant gene: Those genes which are express in F_1.

Recessive gene: Those genes which are not express in F_1.

QUESTIONS

Q. 1. Define Genetics. Enlist different branches of Genetics.

Q. 2. Enlist and explain different principles which are given by G. J. Mendel.

Q. 3. Define genome evolution. Give in brief about evolutionary genetics.

Q. 4. Enlist different characters which are studied by G. J. Mendel.

Chapter 11

Seed Germination

INTRODUCTION

Germination is usually the growth of a plant contained within a seed; it results in the formation of the seedling, it is also the process of reactivation of metabolic machinery of the seed resulting in the emergence of radicle and plumule. It is the process by which an organism grows from a seed or similar structure. The most common example of germination is the sprouting of a seedling from a seed of an angiosperm or gymnosperm.

SEED

It is the fertilized, matured ovule of a flowering plant, containing an embryo or rudimentary plant.

Seed the fertilized and mature ovule containing the embryo is called the seed.

SEED GERMINATION

Seed Germination is the process of reactivation of metabolic machinery of the seed resulting in the emergence of radicle and plumule.

STRUCTURE OF MONOCOT SEED

- The single cotyledon, known as the scutellum, absorbs food reserves from the endosperm.
- The coleoptile and coleorhiza are sheaths that enclose the apical meristem of the shoot and root respectively. After the root emerges the seed sends out its coleoptile and finally pair of true leaves (plumules) emerges.

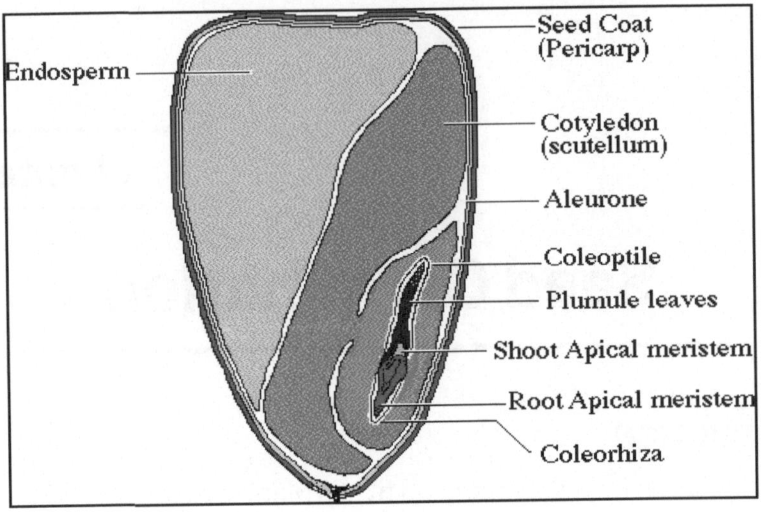

Fig. 11.1. Monocot seed

- The plumules are protected by a sheath of cells called the coleoptile, while the radicle is covered by a similar structure called the coleorhiza.
- The monocot cotyledon - scutellum is used for food storage and digestion. It does not absorb nutrients from the endosperm.

TYPES OF GERMINATION

1. Hypogeal Germination:

In this kind of germination, the cotyledons do not come out of the soil surface. In such seeds the epicotyl (i.e., part of embryonic axis between plumule and cotyledons) elongates pushing the plumule out of the soil. All monocotyledons show hypogeal germination. Among dicotyledons, gram, pea, groundnut are some common examples of hypogeal germination.

In monocotyledons (e.g., wheat, maize, rice, coconut) radicle and plumule come out by piercing the coleorhiza and coleoptile respectively. The plumule grows upward and the first leaf comes out of the coleoptile. The radicle forms the primary root which is soon replaced by many fibrous roots.

2. Epigeal Germination:

In seeds with epigeal germination, the cotyledons are brought above the soil due to elongation of the hypocotyl. In castor, cotton, papaya, onion flat green leaf like cotyledons in the young seedlings. Here the cotyledons, besides food storage, also perform photosynthesis till the seedling becomes independent. In some other plants like bean, the cotyledons being thick do not become leaf-like; they shrivel and fall off after their food reserves are consumed by the seedling.

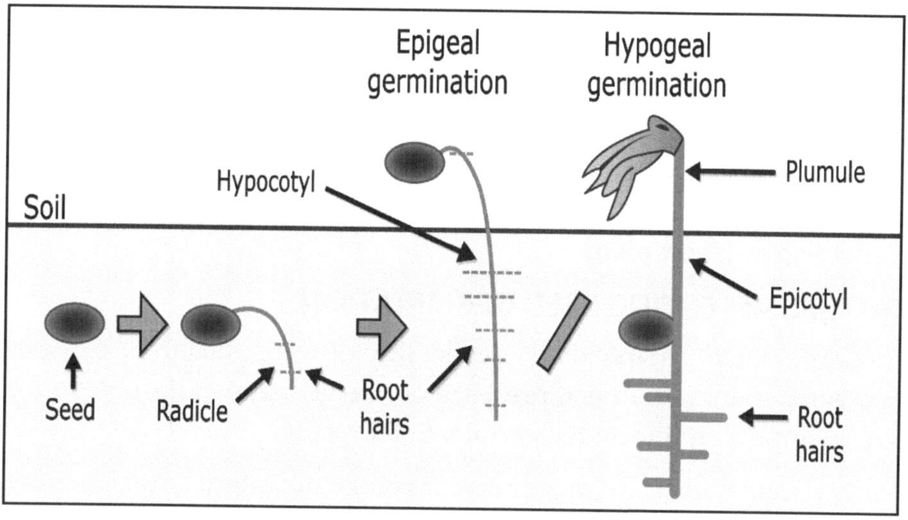

Fig. 11.2. Epigeal and Hypogeal Germination

3. Vivipary (Viviparous Germination):

Vivipary is the phenomenon of giving birth to young ones in advanced stage of development. It occurs in mammals (among animals) and mangrove plants. In mangrove plants (e.g., *Rhizophora, Sonneratia, Heritiera*) the seeds cannot germinate on the ground because of the excessive salt content and lack of oxygen in marshy habitat. In such plants seed dormancy is absent.

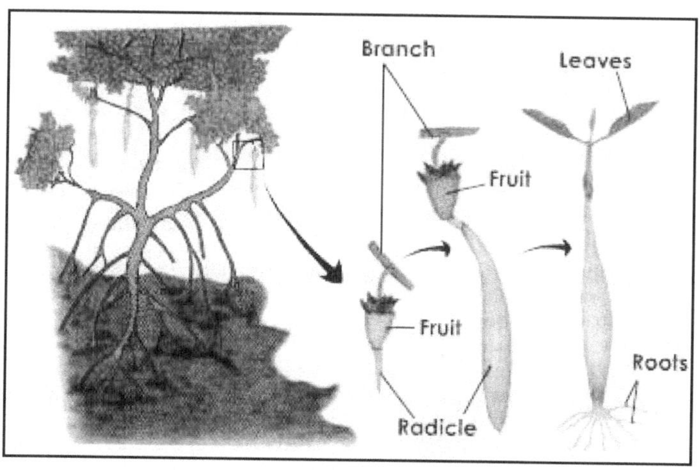

Fig. 11.3. Viviparous Germination

The embryo of the seed (present inside the fruit) continues growth while the latter is attached to the parent plant. Hypocotyl elongates and pushes the radicle out of the seed and the fruit. Growth continues till the hypocotyl and radicle become several centimetres long (more than 70 cm in *Rhizophora*). The seedling becomes heavy.

As a result it breaks its connection with the fruit and falls down in the salt rich muddy water in such a position that the plumule remains outside the saltish water while the tip of the radicle gets fixed in the mud. This protects the plumule. The radicle quickly forms new roots and establishes the seeding as a new plant.

FACTORS AFFECTING SEED GERMINATION

Germination is depends on availability of water, health of the seeds, time, temperature and heat, presence of oxygen, and exposures to light play an important role in the germination process.

- **Temperature:** Extremely low or cold temperature is not favourable for seed germination. They prefer higher temperatures. The germination rate of seed is directly proportional to the rise in temperature.
- **Moisture or water:** Dry seeds do not germinate. Water is an essential factor to trigger off the process of seed germination.
- **Soil:** During growth, seeds require mineral elements for further growth which is obtained from the soil.
- **Light:** For seed germination light is not essential in the early stages of germination but plays a main role in the later stages of the life cycle of plants.
- **Viability of the seeds:** After the seeds are formed, they remain viable up to certain period which may vary from plant to plant or seed to seed. Many sees die or incapable of supporting growth after a certain period of time.
- **Dormancy period:** Many seeds do not germinate abruptly after they are produced. Certain seeds undergoes a resting time through which they stay dormant and germinate when conditions are favourable. Presence of growth inhibitors like abscisic acid induces dormancy in seeds.

- **Thinness or thickness of seed coat:** Different seeds have varying degrees of thickness to enable the seeds to remain feasible. Seeds with a thin seed coat tend to germinate faster than those with thicker seed coats.

GLOSSARY

Seed: It is the fertilized and mature ovule containing the embryo.

Seed germination: It is the resumption of active growth of an embryo of a seed after a period of dormancy.

Epigeal germination: Germination in which cotyledons comes above the soil surface.

Hypogeal germination: Germination in which cotyledons remains below the soil.

Viviparous germination: Seed germinate inside fruit which is still attached to mother plant or seed germination before its dispersal.

QUESTIONS

Q. 1. Define seed. Give the components of the seed with neat labelled diagram.

Q. 2. What do you mean by seed germination? Enlist and explain types of seed germination with suitable examples and neat labelled diagrams.

Q. 3. Define seed germination. What are the factors affecting on seed germination?

Chapter 12

Morphology and Anatomy of Flowering Plants

INTRODUCTION

Morphology is the branch of science which deals with the study of external structural of the plants. The flowering plant consists of an axis, root system and shoots system.

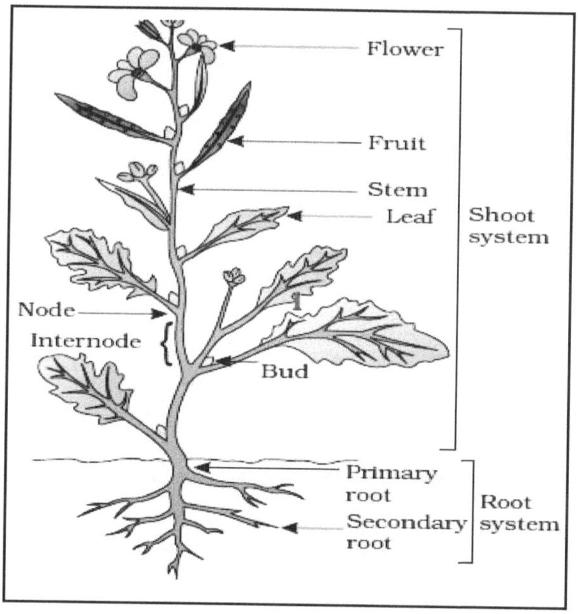

Fig. 12.1. Morphology of flowering plants

ROOT SYSTEM

The flowering plants consist of a long cylindrical axis which is differentiated into an underground root system. Root is the descending, non-green, underground part lacking nodes, internodes, leaves and buds. In Dicotyledonous plants, the elongation of radicle leads to the formation of primary root. The primary root grows inside the soil. Primary root bears lateral roots known as secondary roots, tertiary roots etc.

TYPES OF ROOT SYSTEM

The primary root along with its branches forms a tap root system. Primary roots have short life span and are replaced by large roots later on. For example: Mustard.

The root which arises from the base of the stem is known as fibrous root system. This type of system is common in monocotyledonous plants. For Example: Grasses.

The root which arises from the part other than the radicle is known as adventitious roots. For Example: Banyan Tree, *Monstera.*

REGIONS OF THE ROOT

The root has five main regions which are as follows:

- Root Cap is the structure that covers the apex/tip of the root. It helps in protecting the apex of the root.
- Region of Meristematic activity located few millimeters above the root cap. The cell of this region has the property of repeated divisions which is needed for the growth of the plant.
- Region of Elongation is located proximal to the region of meristematic activity. This is required for the elongation of root.
- Region of Maturation is located proximal to region of elongation. These cells gradually differentiate and mature.
- Root Hairs arise from the region of maturation which helps in absorption of water and minerals from the soil.

MODIFICATIONS OF THE ROOT SYSTEM

Roots modify in structure to perform different functions as explained below:

- Fusiform root is a modified form of tap root. The root is swollen from the middle and tapers at both the ends. For Example: Radish
- Napiform roots are also a modified tap root. They have swollen base and then it tapers abruptly. For Example: Turnip

- ❖ Prop roots are roots which are modified for aerial support. For Example: Corn
- ❖ Stilt roots are modified roots for support. For Example: Maize
- ❖ Pneumatophores are roots modified for respiration. They grow in swampy area and grow vertically upwards. For Example: *Rhizophora*

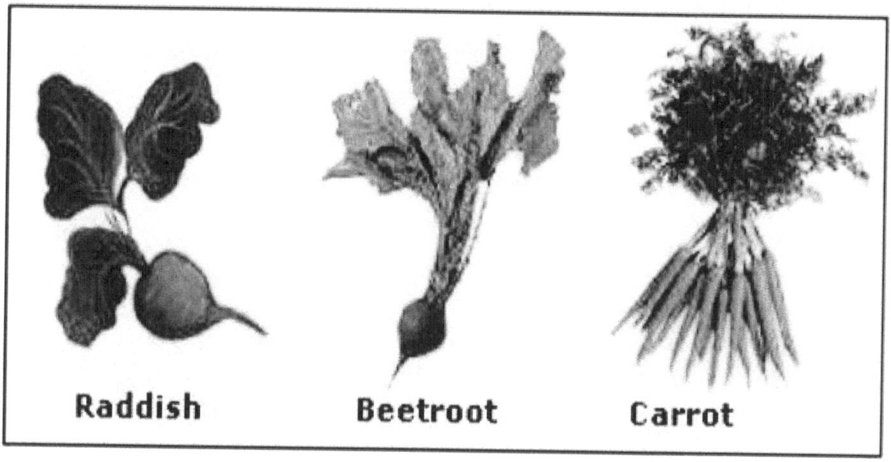

Fig. 12.2. Modification of roots

FUNCTIONS OF THE ROOT

- ❖ To absorb water and minerals from the soil.
- ❖ To provide proper anchorage to the plants.
- ❖ To store reserve food materials
- ❖ Synthesis of plant growth regulators

SHOOT SYSTEM

It is an aerial system, usually above the soil and originates from the plumule. It consists of stem, branches, leaves, flowers and fruits.

STEM

It facilitates conduction of water, mineral and food material. The flowers, fruits and seeds form the reproductive parts of the plants. The stem bears the nodes and internodes. The point from where the leaf arises is known as node and the part between the two nodes are known as internodes.

MODIFICATION OF STEMS

Similar to roots, the stem is also modified to perform different functions:

- Tendrils are slender, twining strands that allow the plant to climb and help in support. For Example: Cucumber.
- Thorns are modified stem for protection of the plant. They are hard, woody and sharp outgrowths from the plant. For Example: Rose.
- Cladodes are modified stem that perform the function of photosynthesis.
- Bulbils are modified stem that becomes fleshy and store food.

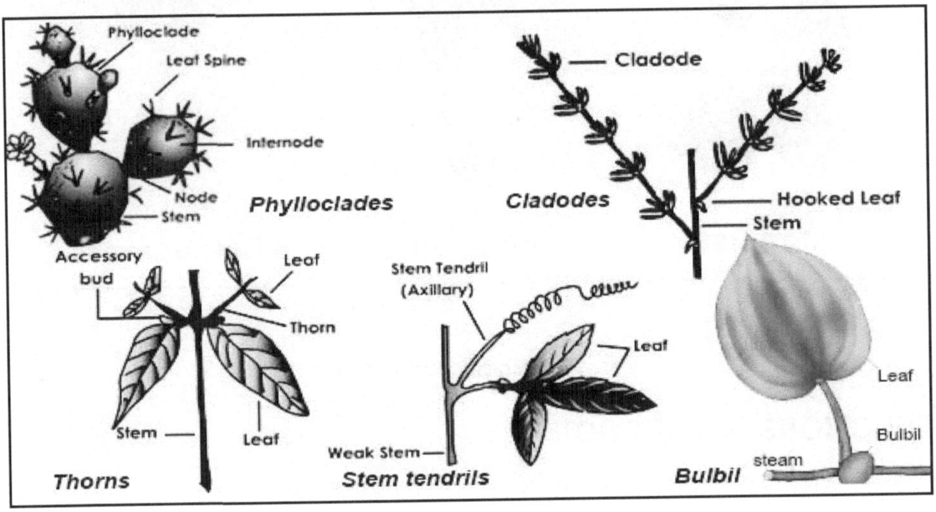

Fig. 12.3. Modification of stem

LEAF

Leaf is a green dorsoventrally flattened exogenous lateral outgrowth that arises from a node of the stem or a branch. The leaf is a specialized organ of photosynthesis, transpiration and gaseous exchange.

PARTS OF LEAF

The point of origin of leaf is known as Node. It bears bud in its axil. The leaves are attached to the stem by the leaf base and have two lateral leaf like structures called Stipules. In monocots, the leaf base is swollen to form pulvinus. The stalk of the leaf is known as Petiole. The green exposed part of the leaf is known as lamina. Lamina bears the veins or veinlets.

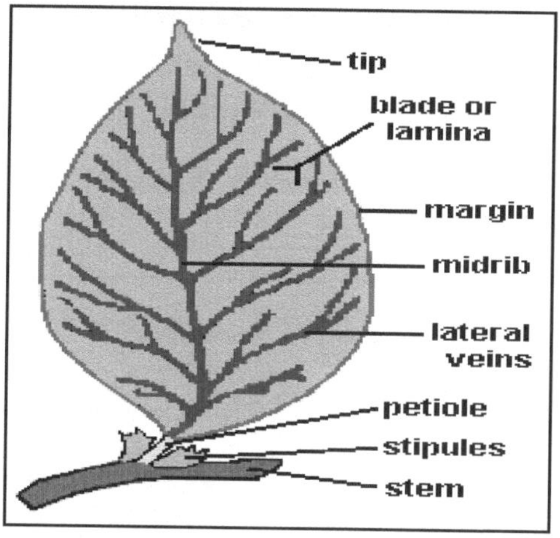

Fig. 12.4. Structure of leaf

STIPULE

The stipules are the small lateral appendages present on either side of the leaf base. They protect the young leaf or leaf primordia. Leaves with stipules are called stipulate and those without them are called exstipulate. The stipules are commonly found in dicotyledons. In some grasses (Monocots) an additional outgrowth is present between leaf base and lamina. It is called ligule. The leaves having ligules are called ligulate. Sometimes, small stipules like outgrowths are found at the base of leaflets of a compound leaf. They are called Stipules.

TYPES OF STIPULES

Depending upon the structure and position various kinds of stipules are recognized.

- Free lateral stipules: A pair of freely arranged stipules present on either side of the leaf base is called free lateral stipules, e.g., *Hibiscus* and Cotton.
- Adnate stipules: The two stipules that fuse with the leaf base or petiole on either side are called adnate stipules, e.g., *Arachis* and Rose.
- Inter petiolar stipules: Stipules present in between the petioles of opposite leaves, e.g., *Ixora* and *Hamelia*.
- Axillary stipules: Stipules present in the axil of a leaf are called axillary stipule. These are also called intrapetiolar stipule, e.g., *Tabernamontana* and *Gardenia*.

- Ochraceous stipules: Membranous tubular stipules that ensheath the axillary bud and a part of internode is called ochraceous stipule. It is formed by the union of two stipules, e.g., *Polygonum* and *Rumex*.
- Hairy stipules: These are hair like stipules which are dry in nature, e.g., *Anacampsora*.

Fig. 12.5. Types of Stipules

LEAF VENATION

The arrangement of veins on the leaf is known as Venation.

- The fine lines present on the leaf are called 'leaf veins'.
- The single middle prominent vein is known as 'mid rib' or the primary vein.
- The veins other than the primary vein are called 'secondary veins'.
- The leaf veins consists of vascular bundles that help in the transport of mineral nutrients and water from the roots to the leaf and the materials produced in the leaf to the rest of the plant.
- The pattern of leaf venation is important characteristic for the identification of plants.
- The veins are present on either side of the mid rib in different fashion.
- The secondary vein can create other pattern where they may or may not be parallel to each other.
- The primary vein can be considered like a 'tree trunk' and the secondary vein as the branches of the tree.

TYPES OF VENATION

- Parallel venation is observed when veins run parallel to each other. For Example: Monocots
- Reticulate venation is observed when veinlets form network. For Example: Dicots

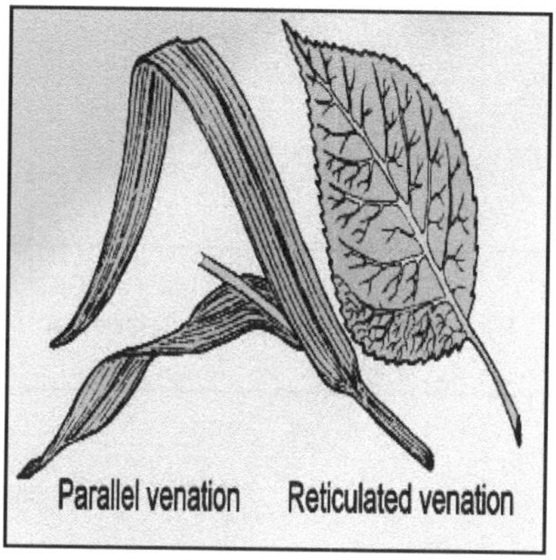

Fig. 12.5. Leaf venation

FLOWER

The modified shoot is known as flower. The arrangement of flower on the floral axis is known as inflorescence. There are two major types of inflorescence- Racemose and Cymose.

- Racemose type of inflorescence: In this type, new flowers are generated at the tip of the inflorescence. For Example: Snapdragon, Mulberry, Oak, Rice, Wheat and Oat.
- Cymose type of inflorescence: In this type, new flowers are generated at the base of the inflorescence. For Example: Cotton, Begonia and Teak.

STRUCTURE OF THE FLOWER

Each flower has four whorls *viz.* sepals, petals, stamens and the carpels. Sepals are a green leafy structure that covers the flower bud. Petals are bright colored to attract insects for pollination. Gynoecium is female reproductive structures whereas Androecium is the male reproductive structure.

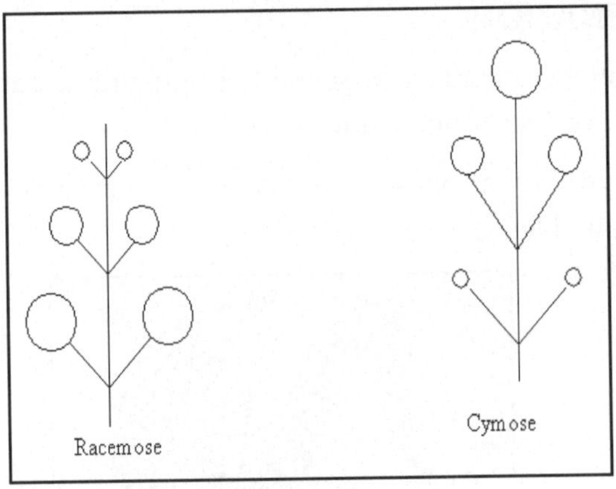

Fig. 12.6. Types of inflorescence

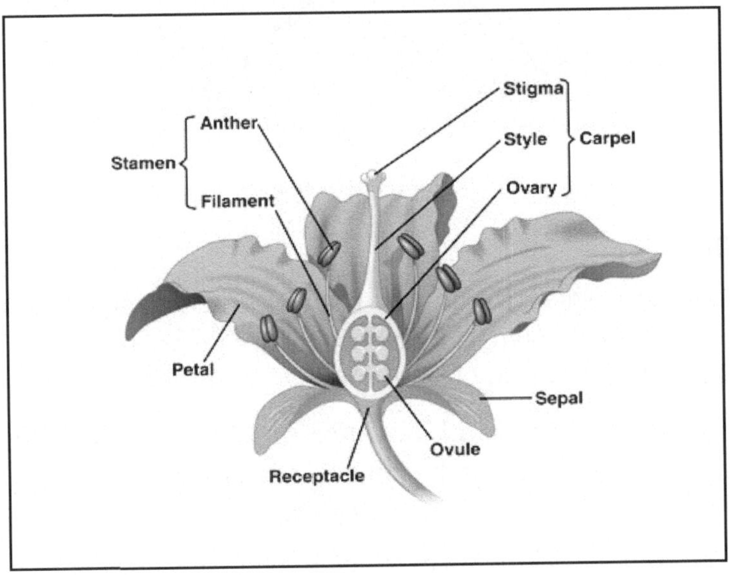

Fig. 12.7. Structure of the Flower

Gynoecium is composed of carpels. Carpels comprises of three structures ovary, style and stigma. Ovary comprises of ovules that forms the seed, whereas stigma is the place of landing of pollen at the time of pollination. Ovary finally forms the fruit. The male part of the flower is known as Androecium. It is composed of stamens. Stamen is composed of anther and the filament. Anthers contain the pollen grains, the male gamete.

AESTIVATION

The arrangement of sepals and petals in bud condition of the flower is called 'aestivation'. It is may be of following types:

- **Open:** If the margins of perianth members in a whorl are free with wide gap between them, then the type of aestivation is called 'open', e.g., sepals of Mustard.
- **Valvate:** Here the edges of perianth members in a whorl are very nearly touching each other bud does not overlap, e.g., calyx and corolla in Annona.
- **Twisted:** In this type, the perianth members of a whorl show one edge outside and one edge inside. Thus they regularly overlap the neighbouring members on one side. The twisted aestivation is also called contorted or convolute aestivation, e.g., corolla of Hibiscus.
- **Imbricate:** Here in a whorl of perianth members, one is completely inside and another is completely outside. The remaining perianth members show one edge inside and the other edge outside. The imbricate aestivation is of two types, namely, descending imbricate and ascending imbricate.

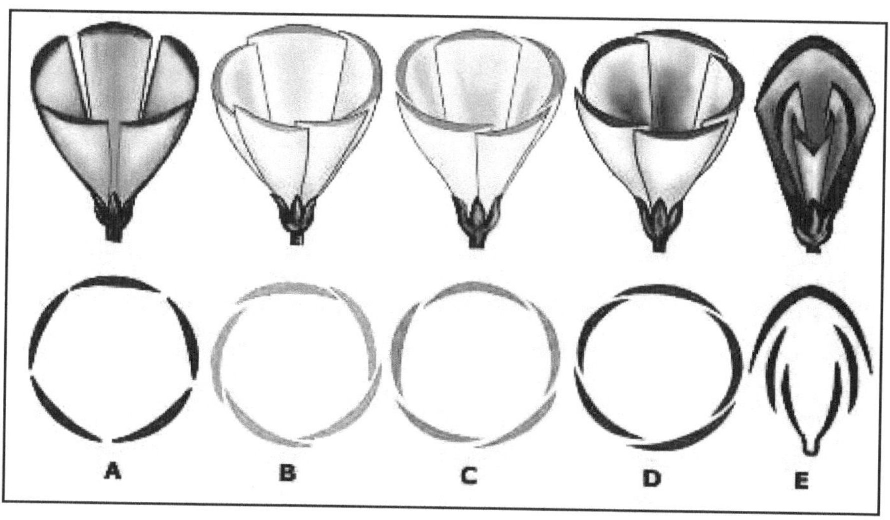

Fig. 12.8. Types of aestivation

- **Descending imbricate:** Here the odd petal is posterior and completely outside. The anterior pair of petals are completely inside.

The remaining petals show regular overlapping in the descending manner. It is also called vexillary aestivation, e.g., *Tephrosia*, *Crotalaria* and *Dolichos*.

- **Ascending imbricate:** Here the odd petal is posterior and completely inside. One of the anterior petals are completely outside. The remaining petals show regular overlapping in ascending manner, e.g., *Cassia* and *Delonix*.

FRUIT

Fruits are the mature ovary or ovaries of one or more flowers. In fleshy fruits, the outer layer (which is often edible) is the pericarp, which is the tissue that develops from the ovary wall of the flower and surrounds the seeds. But in some seemingly pericarp fruits, the edible portion is not derived from the ovary.

1. Pericarp

The pericarp is typically made up of three distinct layers: the epicarp, which is the outermost layer; the mesocarp, which is the middle layer; and the endocarp, which is the inner layer surrounding the ovary or the seeds. In a citrus fruit, the epicarp and mesocarp make up the peel. Exocarp, the outermost layer often consisting of only the epidermis Mesocarp, or middle layer, which varies in thickness Endocarp, which shows considerable variation from one species to another.

2. Epicarp

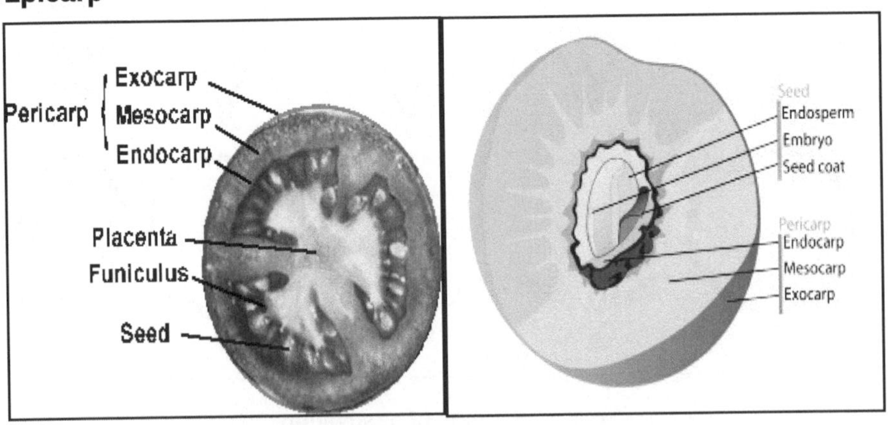

Fig. 12.2. Anatomical Structures of simple fruits

Epicarp (Greek: epi means on or upon and carp means fruit) is a botanical term for the outermost layer of the pericarp (fruit). The epicarp forms the tough outer skin of the fruit, if there is one. The epicarp is sometimes called the exocarp, or, especially in Citrus, the flavedo. Flavedo is mostly composed of cellulosic material but also contains other components, such as essential oils, paraffin waxes, steroidsand triterpenoids, fatty acids, pigments (carotines, chrophylls, flavonoids), bitter principles (limonin) and enzymes.

3. Mesocarp

The mesocarp (Greek: meso means middle and carp means fruit) is the fleshy middle layer of the pericarp of a fruit; it is found between the epicarp and the endocarp. It is usually the part of the fruit that is eaten. For example, the mesocarp makes up most of the edible part of a peach, and a considerable part of a tomato. "Mesocarp" may also refer to any fruit that is fleshy throughout.

In a hesperidium, as is found in citrus fruit, the mesocarp is also referred to as albedo or pith. It is the inner part of the peel and is commonly removed before eating. In citron fruit, where the mesocarp is the most prominent part, it is used to produce succade.

4. Endocarp

Endocarp (Greek: endo means inside and carp means fruit) is a botanical term for the inside layer of the pericarp (fruit), which directly surrounds the seeds. It may be membranous as in citrus where it is the only part consumed, or thick and hard as in the stone fruits of the family Rosaceae such as peaches, cherries, plums and apricots.

In nuts, it is the stony layer that surrounds the kernel of pecans, walnuts etc. and that is removed prior to consumption.

In citrus fruits, the endocarp is separated into sections, which are called segments. These segments are filled with juice vesicles, which contain the juice of the fruit.

GLOSSARY

Leaf venation: It is an arrangement of veins on the leaf.

Stipules: These are the small lateral appendages present on either side of the leaf base.

Inflorescence: The arrangement of flower on the floral axis.

Fruit: It is mature ovary or ovaries of one or more flowers.

Aestivation: It is arrangement of sepals and petals in bud condition of the flower.

Endocarp: It is inside layer of the pericarp of a fruit.

Epicarp: Outermost layer of the pericarp of a fruit.

Mesocarp: Fleshy middle layer of the pericarp of a fruit.

QUESTIONS

Q. 1. What do you mean by morphology of flowering of plant? Enlist and explain types morphological parts of plants.

Q. 2. Define root. What are the types and regions of root? Explain its modification alongwith functions of roots.

Q. 3. Define shoot system. What are the modifications of stem?

Q. 4. What are the parts of leaf? Give the modifications of the leaf.

Q. 5. What do you mean by leaf venation? Enlist and explain types of leaf venation.

Q. 6. What are the components of the flower? Enlist and explain types of inflorescence.

Q. 7. Define aestivation. Explain in brief types of aestivation.

Q. 8. Define fruit. What are the parts of fruit?

Chapter 13

Plant Systematic - Study of Families

A) CRUCIFERAE (BRASSICACEAE)

Table 13.1. Systematic position Cruciferae

Division	Angiospermae
Class	Dicotyledonae
Subclass	Polypetalae
Series	Thalamiflorae
Order	Parietales
Family	Cruciferae (Brassicaceae)

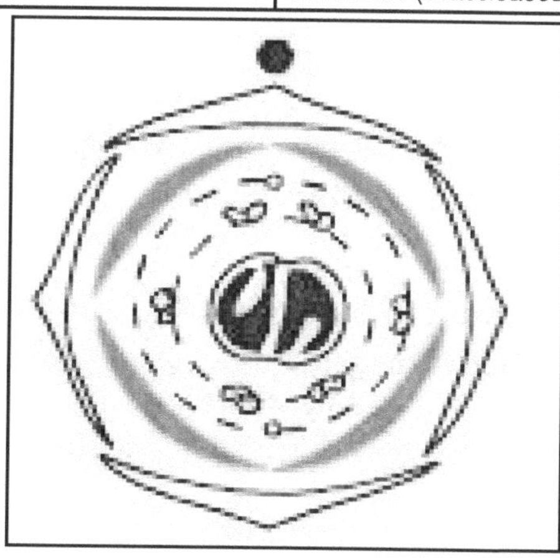

Fig. 13.1. Floral Diagram of Cruciferae Family

- **Habit:** Annual, biennial or perennial herbs. *Farsetia jacquemontii* is an under shrub. The plants possess pungent juice having sulphur-containing glucosides.

- **Root:** Tap root alongwith hypocotyl is swollen in Radish (*Raphanus sativus*) and Turnip (*Brassica rapa*).
- **Stem:** Erect, cylindrical, hairy or glabrous, herbaceous or rarely woody. It is reduced in the vegetative phase in Radish and Turnip. The stem is swollen in Kohlarabi (Knol-Kohl (Ganthgobi, *Brassica, oleracea* var. Gonglylodes). Axillary buds enlarged in Brussel's Sprouts (Button gobhi) or *Brassica oleracea* var. *gemmifera*. *Brassica oleracea* var. *capitata* (Cabbage) has the largest terminal bud.
- **Leaves:** Radical, cauline and ramal, alternate or sub-opposite but forming rosettes when radical, exstipulate with sheathing leaf base, sessile simple or rarely compound (e.g., *Nasturium officinale*), hairy. Bulbils occur in the leaf axils of *Dentaria bulbifera*and on the leaves of *Cardamine pratensis*.
- **Inflorescence:** Flowers are usually arranged in racemose racemes. Occasionally they are in corymbs (candtuft).
- **Flower:** Ebracteate or rarely bracteate (e.g., *Rorippa montana*), pedicellate, complete, perfect, regular, actinomorphic, rarely zygomorphic (e.g., *Iberis, Teesdalia*), tetramerous or bimerous, hypogynous (perigynous in *Lepidium*), cyclic, cruciform.
- **Calyx:** Sepals 4, polysepalous, aestivation imbricate, generally arranged in two whorls, outer of antero-posterior sepals and inner of lateral sepals, lateral sepals generally saccate or pouched at the base, green or petaloid, inferior.
- **Corolla:** Petals 4, polypetalous, arranged in one whorl and alternate with sepals, often with long claws and spread out in the form of a Greek cross. This arrangement of petals which is characteristic of the family is known as the cruciform arrangement and corolla is described as cruciform corolla, valvate aestivation. Petals reduced or absent in *Lepidium* and *Rorippa*.
- **Androecium:** Stamens 6, (four in *Cardamine hirsuta*, two in *Coronopus didymus*, 16 in *Megacarpaea*), free (polyandrous), tetradynamous, arranged in two whorls, outer of two short lateral stamens while the inner whorl is made up of 4 long stamens arranged in two median pairs, anthers basifixed or dorsifixed, dehiscence longitudinal, inferior. Green nectaries are often associated with the bases of stamens.

- **Gynoecium:** Bicarpellary (tricarpellary in species of *Lepidium*, tetracarpellary in *Tetrapoma* and *Tropidocarpum*), syncarpous, carpels placed transversely, ovary superior, placentation parietal, ovary bilocular due to the presence of a false septum called replum, style short, stigma capitate, simple or lobed.
- **Fruit**: Siliqua of silicula, lomentaceous siliqua occurs in radish.
- **Seed:** Non-endospermic, often oily.
- **Floral formula:**

$$\text{Ebr} \oplus ⚥ \ K_{2+2}\ C_4\ A_{2+4}\ \underline{G}_{(2)}$$

B) PAPILIONACEAE (FABACEAE)

Family Fabaceae is large family including plants which are economically important. The family Fabaceae also was known as Leguminosae or Papilionaceae since it is the pea or legume family. It consists of around 5000 species of dicotyledons which are widely distributed all over the world.

Table 13.2. Systematic position Papilionaceae

Division	Angiospermae
Class	Dicotyledonae
Subclass	Gamopetalae
Series	Bicarpellatae
Order	Polimoniales
Family	Solanaceae

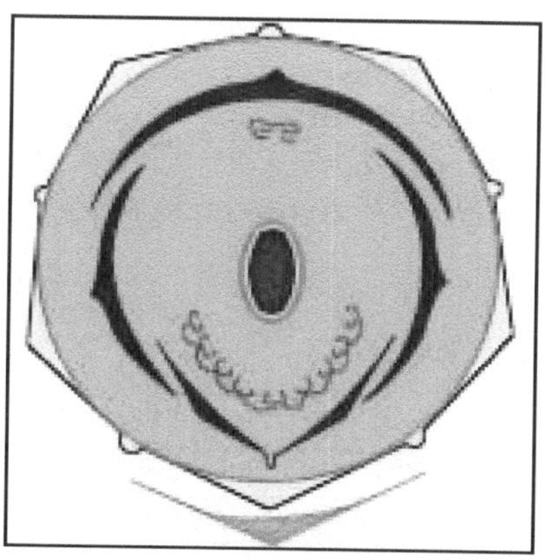

Fig. 13.2. Floral Diagram of Papilionaceae Family

- **Root:** Dicotyledons with taproot with root nodules.

- **Stem:** Erect or climber; Fabaceae include shrubs, herbs, trees and majorly climbers.
- **Leaves:** Petiolate, pinnately compound or simple; pulvinus leaf base, stipulate; reticulate venation.
- **Inflorescence:** Racemose or solitory axillary
- Flower: Bracteate or ebracteate rarely bracteolate (e.g., Arachis), pedicellate or sessile, complete, irregular, zygomorphic, perigynous or occasionally hypogynous and pentamerous.
- **Calyx:** Sepals 5, gamosepalous, usually companulate, lobe unequal, rarely tubular (e.g, *Cyamopsis*), odd sepal anterior, may be persistent inferior.
- **Corolla:** Petals 5, polypetalous, papilionaceous, descending imbricate aestivation, one posterior long standered, two lateral short wings, and two anterior petals jointed to each other forming keel.
- **Androecium:** 10 stamens, usually in diadelphous (9+1 stamens in *Lathyrus* and 5+5 stamens in *Aeschynomene*) or in monadelphous (9 stamens in *Dalbergia*, 10 stamens in *Arachis* and *Erythrina indica*), rarely free (e.g., *Sophora*), nectar gland often present on the inner bases of filaments, anther lobes bilocular, dorsifixed, introse.
- **Gynoecium:** Monocarpellary, ovary superior, unilocular with marginal placentation ovary covered by staminal tube, style bent, stigma simple or capitate.
- **Fruit:** Legume or lomentum.
- **Floral formula:** % ⚥ $K_{(5)}\ C_{1+(2)+2}\ A_{(9)+1}\ \underline{G}_1$

C) POACEAE (GRAMINEAE)

Table 13.3. Systematic position Poaceae

Division	Magnoliophyta
Class	Monocotyledonae
Subclass	Commelinidae
Series	Monocarpellatae
Order	Poales
Family	Poaceae

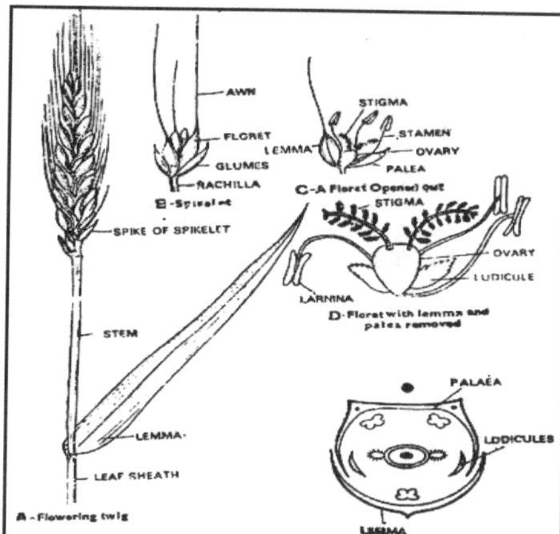

Fig. 118.1. *Triticum aestivum*.

Floral Diagram of Poaceae Family

- **Habit:** Herbs, annuals or perennials or shrubs, sometimes tree like (*Bambusa, Dendrocalamus*).
- **Root:** Adventitious, fibrous, branched, fascicled or stilt (*Zea mays*).
- Stem: Underground rhizome in all perennial grasses, cylindrical, culm with conspicuous nodes and internodes, internodes hollow, herbaceous or woody, glabrous or glaucous, vegetative shoots are arising from the base of aerial stem or from underground stems are called tillers.
- **Leaves:** Alternate, simple, distichous, exstipulate, sessile, ligulate (absent in *Echinochloa*), leaf base forming tubular sheath, sheath open, surrounding internode incompletely, ligule is present at the junction of the lamina and sheath, entire, hairy or rough, linear, parallel venation.
- **Inflorescence:** Compound spike which may be sessile or stalked. Each unit of inflorescence is spikelet. The spikelets are arranged in various ways on the main axis called rachilla. A compound inflorescence may be spike of spikelets (*Triticum*), panicle of spikelets (*Avena*).

The spikelet consists of a short axis called rachilla on which 1 to many sessile or short stalked flowers are borne. The florets may be arranged in alternate or opposite manner on the central axis.

At the base of rachilla two sterile scales, called glumes, are present. The glumes are placed one above the other on opposite sides. The lower one is called first glume and the upper is called second glume. Both the glumes are boat shaped and sterile. Above the glumes a series of florets are present. Each floret has an inferior palea or lemma and above it a superior palea. The lemma frequently bears a long, stiff hair called awn.

- **Flower:** Bracteate and bracteolate, sessile, incomplete, hermaphrodite, or unisexual (*Zea mays*), irregular, zygomorphic, hypogynous, cyclic.
- **Perianth:** Represented by membranous scales called the lodicules. The lodicules are situated above and opposite the superior palea or may be absent, or many (*Ochlandra*), or 2 or 3.
- **Androecium:** Usually stamens 3, rarely 6 (*Bambusa, Oryza*) and one in various species of Anrostis, Lepturus; polyandrous, filaments long, anthers dithecous, versatile, linear, extrorse; pollen grains dry.
- **Gynoecium:** Monocarpellary, according to some authors carpels 3, of which 2 are abortive, ovary superior, unilocular with single ovule, basal placentation, style short or absent; stigmas two feathery or papillate and branched.
- **Fruit:** Caryopsis (achene with pericarp completely united or adherent with the seed coat) or rarely nut (*Dendrocalamus*) or berry (*Bambusa*).
- **Seed:** Endospermic and containing a single cotyledon called scutellum, which is shield shaped and pressed against the endosperm.
- **Floral formula:**

Floral formula – o|o ⚥ $P_{0\ \text{or}\ 2\ \text{(Lodicules)}}\ A_{3\ \text{or}\ 6}\ G_{\underline{1}}$.

GLOSSARY

Floral biology: It is an area of ecological research that studies the evolutionary factors that have moulded the structures, behaviour and physiological aspects involved in the flowering of plants.

Floral diagram: It is a graphic representation of flower structure.

Floral formula: It is a means to represent the structure of a flower using numbers, letters and various symbols, presenting substantial information about the flower in a compact form.

QUESTIONS

Q. 1. Write down detail about Cruciferae family.

Q. 2. Write down detail about Papilionaceae family.

Q. 3. Write down detail about Poaceae family.

Biblography

Adl, S. M.; Simpson, A. G. B.; Lane, C. E.; Lukes, J.; Bass, D.; Bowser, S. S., 2015. The revised classification of eukaryotes. *Journal of Eukaryotic Microbiology*. **59** (5) PP: 429-493.

Avalier-Smith, T. 1998. A revised six-kingdom system of life. *Biological Reviews*. **73** (03) PP: 203-266.

Bernstein M 2006. Prebiotic materials from on and off the early Earth. *Philos Trans R Soc Lond B Biol Sci*. **361** (1) PP: 689-700.

Chatton, E. 1925. Pansporella perplexa. Reflexions sur la biologie et la phylogenie des protozoaires. Annales des Sciences Naturelles - Zoologie et Biologie Animale. **10** (VII) PP: 1-84.

Copeland, H. 1938. The kingdoms of organisms. Quarterly Review of Biology. **13** PP: 383-420.

Datta, Subhash Chandra 1988. Systematic Botany (4th ed.). New Age International, New Delhi.

Griffiths, Anthony J. F.; Miller, Jeffrey H.; Suzuki, David T.; Lewontin, Richard C.; Gelbart, eds. 2000. Genetics and the Organism: Introduction. An Introduction to Genetic Analysis (7th ed.). New York: W. H. Freeman.

Haeckel, E. 1866. Generelle Morphologie der Organismen. Reimer, Berlin.

Hooke, Robert 1665. Micrographia, London, England: Royal Society of London. PP: 113.

Iitians, www.askiitians.com

Linnaeus, C. 1735. Systema naturae, sive regna tria naturae systematice proposita per classes, ordines, genera, & species. Haak, Leiden.

Linnaeus, C. 1753. Species Plantarum. Stockholm, Sweden.

Linnaeus, C. 1758. Systema naturae, sive regna tria naturae systematice proposita per classes, ordines, genera, & species, 10th Edition. Haak, Leiden.

Lloyd, David G.; Barrett, Spencer C. H., eds. 1996. Floral Biology. Studies on Floral Evolution in Animal-Pollinated Plants.

Luketa, S. 2012. New views on the megaclassification of life. *Protistology*. **7** (4) PP: 218-237.

Maton A, Hopkins JJ, LaHart S, Quon Warner D, Wright M, Jill D., 1997. Cells: Building Blocks of Life. New Jersey: Prentice Hall. PP: 70-74.

Matthew Hamilton, 2011. Population Genetics. Georgetown University. PP: 26.

Mayr, E. 1982. The Growth of Biological Thought. Belknap P. of Harvard U.P., Cambridge (Mass.).

Mayr, Ernst 1968. The Role of Systematics in Biology: The study of all aspects of the diversity of life is one of the most important concerns in biology, *Science*, **159** (3815) PP: 595-599.

Paterlini, Marta 2007. There shall be order. The legacy of Linnaeus in the age of molecular biology. *EMBO Reports*. **8** (9) PP: 814-816.

Pearson, H., 2006. Genetics: what is a gene? *Nature*. **441** (7092) PP: 398-401.

PH Raven, Evert RF, Eichhorm SE, 1999. Biology of Plants, 6th edition. WH Freeman, New York.

Prenner, Gerhard; Richard M. Bateman; Paula J. Rudall, 2010. Floral formulae updated for routine inclusion in formal taxonomic descriptions. *Taxon*. **59** (1) PP: 241-250.

Raven, Peter H.; Ray F. Evert; Susan E. Eichhorn, 2005. Biology of Plants, 7th Edition. New York: W.H. Freeman and Company Publishers. PP: 504-508.

Ruggiero, M. A.; Gordon, D. P.; Orrell, T. M.; Bailly, N.; Bourgoin, T.; Brusca, R. C. 2015. A higher level classification of all living organisms. *PLOS One*. **10** (4): e0119248.

Small, Ernest 1989. Systematics of Biological Systematics (Taxonomy of Taxonomy). *Taxon*. **38** (3) PP: 335-356.

Stace, Clive A. 1989. Plant taxonomy and biosystematics (2nd ed.). Cambridge University Press, Cambridge.

Weiling, F., 1991. Historical study: Johann Gregor Mendel 1822-1884. *American Journal of Medical Genetics.* **40** (1) PP: 1-25.

Whittaker, R. H. 1969. New concepts of kingdoms of organisms. *Science.* **163** (3863) PP: 150-160.

Woese, C.; Kandler, O.; Wheelis, M. 1990. Towards a natural system of organisms: proposal for the domains Archaea, Bacteria, and Eucarya. *Proceedings of the National Academy of Sciences of the United States of America.* **87** (12) PP: 4576-4579.

9 789359 190990

Printed at Repro India Ltd.